Information Technology and Innovation in Language Education

Hong Kong University Press thanks Xu Bing for writing the Press's name in his Square Word Calligraphy for the covers of its books. For further information, see p. iv.

Information Technology and Innovation in Language Education

Edited by Chris Davison

香港大學出版社
HONG KONG UNIVERSITY PRESS

Hong Kong University Press
14/F Hing Wai Centre
7 Tin Wan Praya Road
Aberdeen
Hong Kong

ISBN 962 209 696 4

Secure On-line Ordering
http://www.hkupress.org

British Library Cataloguing-in-Publication Data
A catalogue record for this book is available from the British Library.

Printed and bound by Kings Time Printing Press Ltd., in Hong Kong, China

Hong Kong University Press is honoured that Xu Bing, whose art explores the complex themes of language across cultures, has written the Press's name in his Square Word Calligraphy. This signals our commitment to cross-cultural thinking and the distinctive nature of our English-language books published in China.

"At first glance, Square Word Calligraphy appears to be nothing more unusual than Chinese characters, but in fact it is a new way of rendering English words in the format of a square so they resemble Chinese characters. Chinese viewers expect to be able to read Square Word Calligraphy but cannot. Western viewers, however are surprised to find they can read it. Delight erupts when meaning is unexpectedly revealed."

— *Britta Erickson, The Art of Xu Bing*

Table of Contents

Foreword

Much of the discussion about the role of information and communication technology (ICT) in education has focused on whether and how ICTs have been, or can be, fully exploited to enhance the quality of learning. This is a big step forward from the early days when any use of ICTs in the classroom was hailed as innovative teaching, despite merely perpetuating existing practices. Researchers have repeatedly cautioned that new technologies are *merely* tools that may or may not bring about better learning. While caution is prudent, there is a danger that the power of new technologies is being seriously underestimated. Peter Drucker, who at the age of 95, has been described as a seminal thinker with "still the youngest mind" (1997, *Forbes*), has made the following insightful observation:

> We — and I include myself — overrated the possibilities to the point where we talked of computer-generated "business models" that could make decisions and might even be able to run much of the business. But we also grossly underrated the new tools; we saw in them the means to do better what executives were already doing to manage their organizations. (*The Profession of Management,* 2000, p. 83)

Drucker points out that there is a mutually interdependent and interactive relationship between tools and concepts. One changes the other. He elaborates, "The new tools enable us — indeed, may force us — to see businesses differently" (ibid.). Peter Drucker's observation about business is applicable to education. Indeed, new forms of communication, new ways of managing knowledge resources, new relationships among hitherto unrelated groups and new connections among apparently discrete domains brought about by the ubiquitous new information technologies have led to questions about concepts such as "knowledge" and "information" — concepts that have been taken for granted for a long time. As Roy Pea and John Seely Brown point out in the introduction to their thought-provoking series on learning (*Learning in Doing*), computational media are radically transforming the basic patterns of communication and knowledge interchange and by doing so, they are redefining what it means to know and understand, and what it means to become literate or an educated citizen. Seen in this light, when we examine innovation in education, we need to investigate not just how existing pedagogical practices can be improved by new technologies, but more importantly, the affordances for reconceptualizing, renegotiating and redefining learning and knowing that are mediated by new technologies.

In the field of language education, there is still plenty of room for exploring the mutually constitutive relationship between tools and concepts, and between new technologies and learning. This edited volume is therefore an extremely valuable addition to the growing literature on information technologies in language education. The editor, Chris Davison, brings together the work of leading researchers who have been exploring the potential of ICTs in a range of contexts. In an excellent introduction, she gives the volume coherence by locating their work in the contexts of ICTs as innovation and as application, and by outlining common issues germane to the relationship between ICTs and learning. Both practitioners and researchers will find this volume intellectually stimulating and practically relevant.

Amy B.M. Tsui
The University of Hong Kong

About the Contributors

Bob Adamson is a Senior Lecturer in TESOL at Queensland University of Technology. He has published in the fields of higher education, English language teaching, curriculum change and teacher education in the People's Republic of China and Hong Kong. His recent books include *Higher Education in Post-Mao China* (with Michael Agelasto), *Changing the Curriculum* (with Tammy Kwan and K.K. Chan) and *China's English*.

Quentin Allan currently directs an educational consultancy based in New Zealand. In Hong Kong, he worked variously as an English teacher, materials writer, newspaper columnist and teacher trainer. While at the University of Hong Kong, he co-authored *TeleGram* and *PrimeGram*, co-ordinated the *TeleNex* grammar database development team and was responsible for corpus development within TELEC. He has published primarily in the areas of corpus linguistics and asynchronous learning networks.

Phil Benson teaches in the English Centre at the University of Hong Kong. He was formerly co-ordinator of the Centre's Information Technology Integration Project and is active in the development of online courses and learning materials. He is co-editor of *Autonomy and Independence in Language Learning* (Longman, 1997) and author of *Teaching and Researching Autonomy in Language Learning* (Longman, 2001).

Chris Davison is Associate Professor in English Language Education at the University of Hong Kong. She has worked for many years in TESOL teacher education and has published widely in the area of innovation and change in English language teaching. Her recent books include *English as a Second Language in the Mainstream: Teaching, Learning and Identity* (Longman Pearson, 2001) and the two-volume *International Handbook of English Language Teaching* (Kluwer/Springer, in press) with Jim Cummins.

Chris Corbel is General Manager Global English Services at ACL Pty Ltd Sydney. Chris has taught courses at Melbourne and Macquarie Universities on computer literacies, curriculum design and CALL. He is also an Honorary Research Fellow in the Faculty of Education at the University of Melbourne. He publishes regularly on the impact of technology on the work of English language teachers, and has managed several large technology innovation projects.

Jim Cummins teaches in the Department of Curriculum, Teaching and Learning, of the Ontario Institute for Studies in Education of the University of Toronto. His research has focused on the nature of language proficiency and second language acquisition, with particular emphasis on the social and educational barriers that limit academic success for culturally diverse students. His publications include *Language, Power and Pedagogy: Bilingual Children in the Crossfire* (Multilingual Matters, 2000).

Kristina Love lectures in the Department of Language, Literacy and Arts Education at the University of Melbourne. Her research interests include analysis of the spoken and written discourses of secondary English and applications of computer technologies in literacy contexts. Her CD-ROM, *Building Understandings in Literacy and Teaching* (BUILT), uses QuickTime video of best teaching practice to illustrate key principles of language and learning in pre-service and teacher professional development.

Michael K. Legutke is Professor of TEFL at the University of Giessen in Germany. He has worked at the Goethe-Institut Munich, as German Language Consultant in the Pacific Northwest of the US and as language teacher in Germany. His research has focused on language learning in the elementary school, teacher education, and on computer-based learning environments. Among his publications is *Process and Experience in the Language Classroom,* co-authored with Howard Thomas (Longman 1991).

Numa Markee is Associate Professor of English as an International Language at the University of Illinois at Urbana-Champaign. His first book, *Managing Curricular Innovation* (Cambridge University Press 1997), analyzes curriculum design from a diffusion of innovations perspective on social change. His second book, *Conversation Analysis* (Lawrence Erlbaum 2000), investigates how the techniques of conversation analysis may be used to analyze instructed SLA data.

Denise E. Murray is Director of NCELTR and Professor of Applied Linguistics at Macquarie University, Sydney. She was Chair of Linguistics and Language Development at San José State University, California, for nine years. Her research and publications focus on cross-cultural literacy, language education policy, and the intersection of language, society and technology. She was on the TESOL Board of Directors for seven years, serving as President in 1996-97.

David Nunan is Chair Professor of Applied Linguistics and Director of the English Centre at the University of Hong Kong. He has written over 100 books and articles in the areas of curriculum development, teacher education and classroom-based research, his most recent books being (with Ron Carter) *The Cambridge Guide to Teaching English to Speakers of Other Languages* (Cambridge University Press) and *Practical English Language Teaching* (McGraw-Hill, 2004).

Lillian Wong teaches at the English Centre, the University of Hong Kong. As the Assistant IT Integration Project Co-ordinator, she has supported and advised on various IT-related projects. She has extensive experience teaching, co-ordinating and developing web-based teacher training courses. She presents regularly at international conferences and conducts workshops at schools to help teachers implement IT into their teaching. Her current study is on teacher education in IT and ELT.

Acknowledgements

This book had its genesis in the International Language in Education (ILEC) Conference held at the University of Hong Kong in December 2000, on the theme of Innovation and Language Education. The papers by Benson, Corbel, Legutke, Love, and Markee were presented in their original form at that conference but have been substantially revised and updated for this volume. The remaining papers were solicited specifically for this collection. I would like to thank my colleagues in the Faculty of Education, the University of Hong Kong, and other members of the ILEC English Program Committee, without whose encouragement and support this volume would not have been possible. I would also like to thank Eila Thomas whose assistance was invaluable in completing the final editing, Xinmin Zheng who so carefully prepared the index, and Clara Ho, from Hong Kong University Press, whose patience and gentle guidance was most appreciated. I am most grateful to the two anonymous reviewers and to Bob Fox for their helpful comments on the earlier drafts. I would also like to acknowledge the Centre for Information Technology in Education (CITE) for their generous donation of the photographs for the front cover. Finally, I would like to thank my family — Christof, Sean and Nicholas — who are, as always, a constant source of love and support.

Acknowledgements

SECTION I

Introduction

1 Information Technology and Innovation in Language Education

Chris Davison

INTRODUCTION

This book, an edited collection of articles by leading researchers and thinkers in the field, presents a critical analysis and investigation of a range of current developments and debates in the use of information technology (IT) in English language teaching (ELT) internationally. The term *information technology* is used interchangeably with *information and communications technologies* (ICT) to encompass the whole range of existing and potential information and communications technologies used in schools and the wider community, including computers, personal communication devices, digital video and sound systems, email, the Internet and so on.[1] The book is timely for several reasons: firstly, demand for IT use in ELT is extremely high and increasing rapidly worldwide; and secondly, in many countries, policy issues related to the use of IT in ELT, such as its role in learning and teaching and the place of IT knowledge and skills in teacher preparation and credentialing, are controversial and unresolved. ICT is too often seen as either the panacea that will cure all that is dysfunctional or failing in education systems, or as an alien technology that threatens to undermine not just the quality of traditional teaching and learning but many teachers' professionalism and sense of self-worth. In English language teaching, these muddled attitudes and expectations are further complicated by the desire to link distant communities of English language learners and to maximize opportunities to use English for authentic and meaningful purposes.

The reality is that the use of ICTs, like the use of English, is first and foremost a social practice. Hence, to understand the rapidly evolving and changing uses and potentialities of IT in English language teaching, what is needed is an overt socio-cultural and critical analysis of the roles and applications of IT in English language teaching. This book offers such an analysis from a variety of theoretical and geographical perspectives, providing a critical exploration of the variety of ways in which IT is used, misused and could be used in ELT, as a medium of innovation and change, as a "tool" for learning and teaching and as a stimulus for professional development.

The next eleven chapters are divided into three main sections, which focus on these broad themes. The first set of chapters provides an overview of the key issues in IT and innovation in English language education, such as the complex nature of IT and its use in ELT, both in the present and future, and the often-problematic nature of innovation in relation to IT and ELT. It focuses primarily on IT and innovation at the level of programs and curricula, looking at the way organizations and educational systems in different countries, including the US, Australia and the People's Republic of China (PRC), respond to the so-called "IT imperative." The second group of chapters adopts a more overtly social constructivist perspective to explore some examples of innovative practice in IT use in English language learning and teaching around the world. It focuses primarily on IT and innovation at the level of classrooms and pedagogy, looking at the way learners and teachers can and do use IT in their everyday practice. Examples include the role of IT in enhancing the development of second/foreign language learning in Canada, in supporting English a foreign language teaching in classrooms in Germany, in generating discussions about literature in senior secondary English classrooms in Australia, and in facilitating the development of autonomy in Hong Kong. Each chapter explores the way in which the nature of communication and learning has been, or potentially can be, enhanced and changed by the use of IT. At the same time, the limitations of IT and its complex relationship to ELT pedagogy are highlighted. The final section focuses on the problems of building a community of professional practice in IT in English language education. It focuses primarily on IT and innovation at the level of professional development and teacher education. In doing so, it demonstrates how the implementation of IT in schools and classrooms can be enhanced through taking into account key aspects of teachers' existing contexts and professional practices. Professional development initiatives explored include the design of web-based professional development materials and in-service programs for English language teachers in Hong Kong schools, and the establishment of IT competencies for language teachers in the US, Australia and Hong Kong. Throughout the book, the authors take a constructive but critical perspective on the use of IT in English language education, often challenging its role in developing learner autonomy, its effectiveness in developing language learning and its capacity to enhance pedagogic practice in the language teaching classroom, at the same time suggesting many effective models and guidelines for good practice.

The purpose of this introduction is to explore some of the key concepts and theoretical frameworks which underpin the main arguments in the book, in particular, the nature of the innovation theories espoused, the views of language and learning, and of learners and teachers, which inform the explorations of IT as pedagogy, and the views of "community" and "practice" which frame the discussions of both student and teacher learning and development. The following sections of this introduction present these concepts and theoretical frameworks through a discussion of their contribution to the understanding of IT as innovation, as pedagogy and as professional development.

IT AS INNOVATION

The nature of IT is constantly evolving; hence, our definitions and terminology have been changing over time, as Murray (p. 27) points out in the second chapter of this book:

> Within and outside education, a variety of different terms are used to describe the new phenomena. Before we can begin to understand how we might shape and be shaped by these new technologies, we need to deconstruct the plethora of terms, knowing as we do, that the way we name a concept will influence the way we perceive it. The metaphors we choose almost always come from previous technologies and so we find ourselves hobbled to the past while we try to imagine the future, and as we know from other settings, metaphors shape our understanding and view of the world.

Murray's chapter, entitled "New Frontiers in Technology and Teaching," explores how technological innovations in themselves do not determine how people will use them, but rather as new technologies enter our societies, we adopt, change and expand their use in ways not necessarily intended by the inventors of the technology. She argues that our ability to adopt and adapt new technology in order to deliver instruction in innovative ways is often impeded by the very language we use to describe the technology. By focusing on *information* technology, Murray argues, we are appropriating terminology from the technical field, taking a view of information that is inappropriate for developments in language education. In using terms like email, we remain mired in metaphors of older technologies, preventing us from exploiting new potentials (p. x, this volume). This tendency is evident in many examples of online teaching in which lecture notes, perhaps supplemented by PowerPoint slides, are often equated with e-learning. Such examples of new technologies that are used to sustain traditional pedagogies can be contrasted with the much more dynamic, socio-cultural and constructivist uses of IT in education, exemplified in the extensive mechanisms for synchronous and/or asynchronous communication available via the Internet. Murray goes on to explore new developments in digital technologies, such as convergence and wireless connection, explaining how these might affect English language teaching and learning and what critical issues might arise. At the same time, she insists that understandings about language learning should drive how we exploit these developments as a tool for improving teaching and learning.

Murray's arguments problematize views of innovation in IT, the main theme explored in the first section of this book. Following Fullan (1991; 2003), many of the contributors make a critical distinction between innovation and change. This distinction has been elaborated in the ELT field by Stoller (1997, 34):

> Change is predictable and inevitable, always resulting an alteration in the status quo but not necessarily in improvements … . Innovation, on the other hand, results from a deliberate and conscious effort that is perceived as new, as intended to bring about improvement, and has potential for diffusion.

Stoller (1997, 45) argues that we should work towards deliberate rather than unplanned efforts to promote innovation; however, the innovation literature demonstrates that planning alone cannot ensure success. The reason why some innovations are so successful but so many others fail is rarely addressed. Carlopio (2003) suggests reasons include both macro-level issues, such as organizational culture and structure, as well as micro-level issues, such as the perceived characteristics of the innovation, and the characteristics of individuals. Rogers (1983), in his well-established diffusion model, suggests that in any innovation, individuals fall into one of five distinct categories (innovators, early adopters, early majority, late majority, and laggards) according to their social and psychological characteristics (p. 243). Moore (1991) redefines these categories from the perspective of a change agent wishing to introduce a technological innovation, identifying three markets for any technological innovation: early, mainstream, and late. Geoghegan (1994, 6–7), elaborating on Rogers (1983) and Moore (1991), describes the five categories of adopters as follows: Innovators (two to three percent of adopters) are as interested in the technology as the application and may be members of broad, cross-disciplinary networks of similar IT-oriented individuals. Early adopters (about ten percent of adopters) are interested in the possible application of the technology to their professional context. They are task-oriented risk takers, happy to experiment, and fairly independent. The early majority is far more pragmatic, less willing to experiment, demanding proven applications. Their networks are vertically discipline-based, rather than cross-disciplinary. The late majority are similar to the early majority, though perhaps less confident with IT. The last fifteen percent, the laggards, may ignore the innovation completely. Hence, within the ELT field, teachers and learners cannot be treated as a homogeneous group; they can include adopters and resistors, innovators and implementers, change agents and the "unchanged."

Furthermore, as the literature on educational innovation demonstrates, change is not a linear or even rationale process, nor are participant roles and reactions fixed and predictable. Lissak (1999, 12) argues:

> Often the emergent nature of change as experienced by other members of the organization in overlooked. Change, instead, is treated as continuous, step-like, or even chaotic but with a definable scope and focus. The experienced sense of change — that the whole is bigger than the sum of the parts and that the patterns observed and felt are unexpected — is not captured.

The problem of managing innovation is exacerbated with IT, as new educational technologies tend to encourage innovation based on intuition and tacit knowledge, with the result that practice in many respects often leads theory, or even contradicts it in unexpected and often surprising ways. As Davison (this volume) points out, if we try too hard to predict the unpredictable, we may undermine the development of the very knowledge, attitudes and values that we are trying to develop in our teachers and learners.

The third and fourth chapters address these issues, including the nature of planned and experienced change in IT and English language teaching. Markee, in "Making Curriculum Innovation Work: The Role of Information Technology Centers," argues that as the World Wide Web and other information technologies become ever more widespread and pervasive in people's everyday lives, there is increasing pressure on educational institutions to use these technologies as resources for pedagogy. He argues that while IT is undoubtedly enormously seductive, its use in educational contexts must be carefully thought out, particularly in the context of second *language* education. In particular, issues of user support must be carefully planned. Markee draws on work on innovation by Fullan (1991) and Rogers (1983) to analyse the *what*, *why* and *who* of innovation in IT and ELT. Drawing on the experience gained by the Language Learning Laboratory at the University of Illinois at Urbana-Champaign, Markee illustrates the kinds of problems that must be addressed in order to make IT-based curricular innovations in second language teaching actually work. He argues that successful IT centers must be closely linked to curriculum initiatives that are important priorities for the wider institution, if substantial investments in IT are to be justified. Markee adds that the issue of ongoing support for end users must be solved if individuals who have no inherent interest in IT are to be persuaded that it is worth their while to use such technologies in the long term.

The diffusion model implies a smooth transition between stages of adoption. But in the fourth chapter, Corbel, drawing on the work of Moore (1991), identifies a potential "chasm" between uptake by the early market— *innovators* and *early adopters* — and the mainstream market — *early* and *late majority* (Moore 1991, 17). Corbel, in his chapter entitled "Building a Bridge: A Case Study in Crossing the Innovation Chasm," addresses the question of how to make IT use an integral component of the practice of all teachers, not just the early innovators. Using the concept of the existing independent learning centers as a metaphor for the operation of a parallel virtual online network, Corbel explores the aims and operation of the Virtual Independent Learning Center in Victoria, Australia. The Virtual ILC is a collection of language learning tasks, created by teachers, based on real websites and other media resources. The tasks focus the students on specific curriculum goals, thus making efficient and effective use of their Internet time. As an innovation in IT, the Virtual ILC is a simple yet very effective model of IT learning and teaching which offers everyone the opportunity to work online, not just the enthusiastic few. It provides a bridge across Moore's chasm, by explicitly addressing Geoghegan's (1994) four necessary conditions; that is, recognizing that the mainstream, the majority of adult ESL teachers in the organization, are distinct and different and not easily brought around to the visions of the innovators and early adopters; ensuring that mainstream support is provided by ESL teacher-mentors rather than by technophiles; offering incentives and rewards for involvement, thus some reason to adopt the innovation; and emphasizing the institution's commitment to IT adoption for all. Even more importantly, however, Corbel demonstrates the importance of building a bridge between what is familiar and known pedagogically and what is new for both students and teachers. This is particularly important for adult English as second language learners who may have had little exposure to IT and little opportunity to use it, and who may not have high levels of education or literacy in their first language. However, it is even more important for the majority of ESL teachers who are juggling a multitude of demands on their time and attention and are already hard pressed to deal effectively with familiar routines and concepts.

This focus on the end user and on the importance of context and situation are taken up even more overtly in the fifth chapter. In "Developing IT for English in Chinese Secondary Schools," Adamson describes a US-Chinese project to develop an innovative IT package in English language teaching to supplement the textbook series that is used in the majority of schools in China. The study focuses on how the often-controversial issues of technological and linguistic transfer, central to the increasingly global use of IT and the English language, were handled by the various stakeholders. An examination of the four critical dimensions of educational policy identified by Elmore and Sykes (1992) — the genesis of the project, the processes of operationalization, the nature of the product, and the impact of the project — shows how the time-honored principle of grafting Western technology onto Chinese realities (formulated in the Qing dynasty) was operationalized. The study concludes that effective international transfer is not a simple donor-recipient relationship; instead, it is a dynamic, creative process of mutual learning, with particular emphasis on the expertise, practices and developmental capacity of the end user. Adamson's arguments lead into the main theme explored in the next section of the book, that of IT as pedagogy.

IT AS PEDAGOGY

As has been implied earlier, IT sets limits on what may be pedagogically possible, but the pedagogy itself is a far more limiting, or liberating, factor in the use of IT in English language education. The limits of IT are exemplified in Murray (this volume), who argues that one of the major constraints in using most current technologies is that they are not available for "just-in-time" learning. Time in the computer lab has to be planned for and inevitably is not available just at the moment the teacher wants to use a webpage or email in her teaching. With wireless devices, whether handheld or laptop size, however, teachers can integrate ICT into their classroom just as they currently integrate the OHP or the whiteboard and chalk. Similarly, if discussion software proves to be supportive of language learning, teachers will be able to use it for asynchronous and synchronous online learning, learning that would include audio and video via broadband networks. Murray adds that with the advent of broadband, we can expect "classroom" materials that are truly realia. Designers will need to design for the different devices — materials that work optimally on a broadband network with high-speed cable to a fixed line will not be optimal for a handheld wireless — materials designers to develop ICT language learning materials that students can comfortably use. IT by itself cannot transform learning, but the provision of reliable and economic information technologies can certainly add to teachers' pedagogic potential.

Such arguments highlight the issue of IT and pedagogy. Cuban (2000, 97–8), in a wide-ranging study of teacher IT practices in North America, found that despite abundant access to IT and contrary to the expectation of promoters, teachers made infrequent and limited use of computers in classrooms, and uptake or lack of it are apparently unaffected by age, experience, gender, or even the provision of customized training. Furthermore, teachers who incorporated some IT use into their classrooms had "adopted an innovation to fit their customary practices, not revolutionize them" (Cuban 2000, 97).

The yet unrealized potential of many of the new technologies focuses attention on the question of the most appropriate pedagogy to harness IT in English language teaching. In the ICT field, as in the English language teaching field, much current innovative pedagogy has been influenced by situated cognition and Vygotskian thought (Vygotsky 1978, 1981) and by the conceptualization of learning though participation in *communities of practice* (Lave and Wenger 1991). As Cummins (p. 105, this volume) elaborates, the Russian psychologist Lev Vygotsky (1962; 1978) was interested in the distance between the levels of development or problem-solving that children were capable of on their own, without adult guidance, and the level of development that children could potentially attain under the influence of, or in collaboration with, more capable adults and peers. Expressed simply, this *zone of proximal development* (ZPD) is "the interpersonal space where minds meet and new understandings can arise through collaborative interaction and inquiry." Vygotsky's work has been extremely influential on the evolution of social constructivist approaches to learning that emphasize the importance of learners' prior experience in the collaborative scaffolding and construction of knowledge.

Similarly, Lave and Wenger's (1991) theory of *legitimate peripheral participation* conceptualizes learning as a process of evolving participation in numerous and overlapping *communities of practice*. Communities of practice are defined as informal social organizations resulting from mutual human engagement in joint enterprises. Over a period of time, participants in these communities adopt common forms of interaction and discourse, beliefs and values, and ways of doing things, as well as shared resources, concepts, and symbols. In contrast to approaches that view learning as an isolated experience in which individuals engage in order to gain specific knowledge and skills, Lave and Wenger emphasize that "learning, thinking, and knowing are relations among people in activity in, with, and arising from the socially and culturally structured world" (1991, 51). By locating learning in the processes of evolving participation in social practice, the focus shifts away from teaching "onto the intricate structuring of a community's learning resources" (Lave and Wenger 1991, 94). These resources include the social relations necessary for the development of identities of belonging; diverse forms of collaboration that enable participants to perform at levels beyond their individual capacity and thus grow into new levels of competence; the community's shared history, tools, and understandings; and the unique knowledge possessed by individual members. Through engaging in increasingly central roles and responsibilities in practice, participants' access to community members, activities, discourses, and technologies expands, and they experience new opportunities to develop knowledge, skills, and identities. In contrast, confining learners to a narrow range of participation limits their access to learning resources and their chances for "absorbing and being absorbed in 'the culture of practice'" (Lave and Wenger 1991, 95) of the community.

In the ICT field, Hung and Chen (2001) derived four principles for learning to guide the design of web-based e-learning environments from an analysis and synthesis of the literature on situated cognition, Vygotskian thought, and communities of practice, underpinned by a view of the social dimensions of IT (Brown, Collins and Duguid 1989).

The four principles are commonality, situatedness, interdependency and infrastructure (Table 1.1).

Table 1.1 Selected Instructional Design Considerations for E-Learning (adapted from Hung and Chen 2001)

Principles of situated cognition and Vygotskian thought	Instructional design considerations for online teaching and learning
Commonality: Learning is a social act leading to identity formation and associated membership of a community of practice.	E-Learning environments should capitalize on social and collaborative communication with others who have shared interests.
Situatedness: Learning is reflective, metacognitive and embedded in rich socio-cultural contexts.	E-Learning environments should enable students to work on activities and projects that demand reflection on authentic practice.
Interdependency: Learning is socially mediated and facilitated through engagement in practice with others.	E-Learning environments should generate interdependencies that benefit from the diverse expertise in the learning community.
Infrastructure: Learning is facilitated by activity, accountability and associated support mechanisms.	E-Learning environments should incorporate facilitating structures, accountability mechanisms, and associated rules of engagement.

From an instructional design perspective, students (or teachers) in the same field are assumed to be social actors who have a shared interest in and identification with the object of study (commonality). Effective learning is assumed to occur through the opportunity to work on authentic projects that demand interaction and reflection (situatedness). Learning takes place through interaction with peers and more capable others; hence, diversity within the cohort is necessary to create the required conditions for enhanced learning from each other (interdependency). "Facilitating structures, accountability mechanisms, and associated rules of engagement" provide the essential infrastructure to enable such teaching and learning to occur. These principles highlight the potential of IT for enhancing ongoing meaningful social engagement among students and teachers, and are consistent with other theorists in ICT such as Brown and Duguid (2000, 127), who stress the importance of conceptualizing learning as a social act: "Practice is an effective teacher, and community of practice an ideal learning environment."

An additional and complementary perspective on IT as pedagogy comes from adopting a socially oriented view of language. If we look at IT as a social resource to create meaning, as in a systemic-functional linguistic model of language learning (Halliday 1981), we can see the potential of the IT medium in new ways. For example, using the ever-increasing resources of the Internet as well as digitalized information, IT can be used to build knowledge of the *field* — the subject matter of the English language texts and their related concepts and vocabulary — in a much more elaborate and individualized way, as Cummins shows in his chapter (this volume). Using asynchronous communication networks, including email and web forums, IT can also expand, even transform, the *tenor* of relationships between teachers and students, and between students themselves, through offering a enormous variety of opportunities to adjust the type, frequency and quantity of the interaction and interactants, as is exemplified by Legutke (this volume). Finally, despite its yet inadequate inability to handle real-time spoken conversation (see Markee, this volume), IT offers significantly increased opportunities to exploit the full *mode* continuum from most spoken-like to most written-like, providing an enormous diversity of text-types, including those in other languages, and communicative opportunities ranging from most context-embedded situations to the most formal unfamiliar situations. At the same time, IT is a mode that, in general, provides opportunities for greater reflection, thereby facilitating learners making connections among ideas and constructing coherent knowledge structures.

As Murray (this volume) explains, ICT has also resulted in new text types, and new structures and visual features are increasingly used to convey information, not just to support printed text, a situation that has been described as "a tectonic shift" (Kress 1997) and "the breakout of the visual" (Bolter 1996). As Kress (2000, 337, cited in Murray, this volume, p. 38) observes, it "is now impossible to make sense of texts, even of their linguistic parts alone, without having a clear idea of what these other features [audio, video, images] might be contributing to the meaning of a text." Murray argues that we need linguistic descriptions of these new text types and modes of communication, and then strategies for teaching these new text types and communication processes, just as we have a variety of approaches to the teaching of written texts such as text and genre-based approaches, and process writing. At the same time, the incorporation of IT into teacher-student and student-student interaction creates new pedagogic challenges, as "merely offering people a network-connected machine will not ensure that they can use the medium to meet their needs because they may not be able to maximally take advantage of all that the (medium) has to offer" (Hargittai 2002, 13).

Not surprisingly, the contributors to this volume have all been influenced to some extent by these more socially oriented constructions of language, language learning and teaching. In the sixth chapter, "Using IT to Create a Zone of Proximal Development for Academic Language Learning: A Critical Perspective on Trends and Possibilities," Cummins (this volume, p. 105) explores the extent to which IT can provide scaffolds to support the development of first and second language (L1 and L2) literacy abilities, a process he refers to as *academic language learning*. He presents a framework for computer-supported academic language learning that views the interactions between educators and students as the most immediate determinant of student success or failure in school:

> These interactions can be viewed through two lenses: the lens of *the teaching-learning relationship* in a narrow sense, represented by the strategies and techniques that teachers use to provide comprehensible input and reading instruction as well as promote content knowledge and cognitive growth; the second lens is the lens of *identity negotiation* that is represented by the messages communicated to students regarding their identities — who they are in the teacher's eyes and who they are capable of becoming.

More generally, the chapter examines the relationships between pedagogy and technology. It addresses questions such as: What pedagogical options are available for use of IT in schools? To what extent can IT amplify the impact of different approaches to pedagogy? Is the impact of IT potentially more powerful as a catalyst and support for learning when harnessed to certain forms of pedagogy as compared to others?

The chapter begins with the presentation of a theoretical framework for academic language development with three foci: focus on *meaning*, focus on *language*, and focus on *use*. The chapter then explores how this framework has been implemented in a computer-supported language learning program entitled *e-Lective Language Learning*. The program is based on the premise that written text can serve as input for the language learning process. Furthermore, the development of academic language proficiency *requires* that students get extensive access to, and be enabled to harvest, the language of academic text. Any text in electronic form (downloaded from the Internet, scanned in, or available on CD-ROM) can be imported into the program. Thus, teachers have the opportunity to select stories and expository texts that match their students' interests and cultural backgrounds rather than relying on one-size-fits-all programs. Students get one-click access to L1 and English dictionary support to facilitate understanding of the meaning of individual words and sentences. The program "remembers" the words that each individual student has clicked (unknown words) and provides individualized practice to assist students in learning this vocabulary. These practice exercises employ several varieties of receptive and productive cloze procedure and can be set at five levels of difficulty.

Cummins (this volume, p. 123) argues that such uses of IT represent a highly effective tool both to challenge coercive relations of power and promote academic language learning among students from marginalized communities, when it is rooted in a *transformative* rather than transmissive pedagogical orientation:

> Instruction can increase cognitive engagement and identity investment when the power of IT is harnessed to connect students with sister-classes in different cultural, linguistic, and geographical contexts. Students' own identities can be expanded when they have to define aspects of their own reality to present to the sister-class and when this reality is reflected back to them through discussion and project collaboration. Literacy is expanded in the process of carrying out joint projects for the simple reason that students read much more and write much more than when they are in an isolated, contained, non-communicative classroom situation.

However, Cummins (this volume, p. 124) argues that such projects rooted in a transformative pedagogy orientation are still extremely rare, despite their great potential to enhance cognitive engagement, knowledge generation, and identity investment:

> It is much safer to remain within a traditional pedagogical orientation where pre-scripted content is transmitted for students to internalize without elaboration or dissent than to conceptualize the zone of

proximal development as a knowledge construction zone where social understandings can be negotiated in the process of critically analyzing academic content.

In the seventh chapter of this volume, "Redesigning the Foreign Language Classroom: A Critical Perspective on IT and Educational Change," Legutke also explores the relationship between IT and pedagogy, arguing that IT needs to be integrated into what teachers do on a day-to-day basis, in order to realize its educational potential as a multifaceted tool and thus help to transform the learning environment. He draws, albeit implicitly, on Lave and Wenger's (1991) theory of legitimate peripheral participation, stressing the importance of access to authentic and collaborative forms of participation and practice for young English as foreign language school learners, including access to such valuable learning resources as carefully structured collaborative activity with native speaking peers, which can also help them to establish viable identity positions as students and speakers of English in the wider, albeit virtual, community.

Legutke takes the notion of task-based and experiential language learning as a starting point for his chapter and expands on the idea of the language classroom as a communication center and a studio for text production. Drawing on data from three case studies of different English as foreign language classrooms in different contexts (young learners in their second year of language study, intermediate, and advanced learners), Legutke discusses the design features and characteristics of such a classroom. All three classrooms use IT, two of them in learning networks with classrooms from other countries. Legutke examines in more detail the "representational resources" used by learners as a means of co-constructing a meaningful learning environment; the role of computer technology in reshaping the space of action available to both teachers and learners, and the relationship of electronic literacies and educational practices within a larger framework of a negotiated curriculum.

In the first case study, IT is "peripheral" in a sense that in this elementary class, the computer is used in its very simple but powerful function as a writing tool and presentation device and as a tool to obtain information on dream trips. In the second case, in a secondary school, the computer is used as a research tool. Here, the learners leave the limited world of the textbook (providing inferior and outdated text material) to explore more complex content through the web. The content of their presentations derives solely from this research. In the third case, the previous functions are utilized in addition to the computer's potential as a tool for intercultural communication. By locating learning within the processes of evolving participation in IT as social practice, Legutke shows how the pedagogic focus shifts away from instruction per se and "onto the intricate structuring of a community's learning resources" (Lave and Wenger 1991, 94).

Both Cummins and Legutke demonstrate in different ways how IT in itself cannot transform ELT classrooms, but embedded within a constructivist pedagogy it can be used by the informed and confident teacher to boost English language learning outcomes significantly. One of the most obvious examples of this effect in English language teaching is the way IT enables the building of powerful interactive networks and connections to information and people around the world (see also, for example, Cummins and Sayers 1995; Kourtis-Kazoullis and Skourtou, in press), which can not only greatly enhance communicative purpose and student motivation but also significantly increase opportunities to use English in a range of authentic and meaningful contexts.

In her chapter entitled "Online Discussion in Secondary English: Shaping Interactive Practice for ESL Students," Love uses a framework derived from Hallidayan systemic functional linguistics to both describe and evaluate one such ICT network, the use of discussion forums as an integral component of ELT pedagogy. In doing so, she breaks new ground by providing a detailed and principled description not only of the pedagogic possibilities of IT but the actual processes of communication and interaction. Her chapter is a report of an evaluation of the effectiveness of online discussions as an approach to responding to literature in the senior secondary English curriculum in a "technology-rich"' school in Australia. There were 160 Form 5 (Year 11) students divided into groups of six, and a teacher moderator who monitored and assessed their contributions. The school wanted to evaluate the pilot for a number of important reasons: there were no established models of online discussions being used for learning purposes in secondary English; the cohort was a large one, preparing for a high-stakes examination; the school was using a web-based CMC program, designed in-house for the specific purposes of the online discussions, and it was important to evaluate the design features of this program and how effectively it met the needs of the teachers and students; and finally, the school was global in outlook and wanted to broaden the use of online discussions to include interaction with other schools around the world. Love outlines some of the findings from this evaluation and discusses their relevance to schools in other contexts that are considering using online discussion for text response purposes. At the same time, she also makes it clear that different participants use the new technology in different ways. The concepts of field, tenor, and mode were most useful in providing a systematic means for examining the social, academic, and organizational outcomes of teachers' pedagogical choices in this new online mode and helped to guide the ESL teacher in her planning for the next stage of her work with the online discussions, centered on a new literary text.

In the final chapter in this section on IT and pedagogy, Benson also explores the ways in which the quality of IT interaction can affect language learning, through a focus on the development of learner autonomy. Many advocates of ICT in language education cite its fostering of learner autonomy as an advantage of IT-based over traditional classroom instruction. While studies have shown that learners can become independent learners (see, e.g., Jones, Valdez, Nowakowski, and Rasmussen 1995), others have shown that learners need IT-based work to be scaffolded (see, e.g., Murray 2003; Love, this volume; Cummins, this volume). In his chapter entitled "Autonomy and Information Technology in the Educational Discourse of the Information Age," Benson discusses several examples of current practices involving IT in language learning in Hong Kong secondary and tertiary systems. He argues that the key issue in evaluating the effectiveness of these practices is not whether the technology itself is innovative but whether it is used in pedagogically innovative ways that foster autonomy. He adopts a very critical perspective on IT, arguing that there is perhaps too much pressure to adapt educational systems to an increasingly technologized world without necessarily evaluating the benefits for our students' learning and our own professional development.

Benson's chapter, in demonstrating how system-level needs are not always the same as classroom-level needs, serves as a timely note of caution about the rush to innovate in IT and establishes the foundations for an exploration of the third major theme of this edited collection, that of IT and professional development.

IT AS PROFESSIONAL DEVELOPMENT

As Murray (this volume) notes, teachers have reacted to the introduction of ICT in education in a number of ways: fear of becoming redundant, resistance to having to learn new ways of teaching, concern about learning the technology, uncertainty about whether the technology will dominate pedagogy, or enthusiasm at the potential of the new technologies to facilitate learner autonomy. Although the fears and worries are understandable, given the research literature on change and innovation (see, e.g., Fullan 1991), Murray argues that IT introduces new complexities into the equation. This final section of the edited collection focuses on the relationship between IT and professional development. The term *professional development* is deliberately ambiguous, as the issues in this section encompass both the need for individual teachers' development in knowledge and skill to make more effective use of IT in learning and teaching, as well as the need for the ELT profession itself to expand its professional practices to make IT an integral and widely used resource in its learning/teaching community.

In their chapter entitled "Innovation and Change: IT and In-service Teacher Education," Nunan and Wong report on an empirical investigation into the effect of an innovative, in-service program on the teaching practices of secondary English teachers in Hong Kong. The purpose of the chapter is to explore the teachers' uses of technology, their perceptions of the value and role of technology in the classroom, and their beliefs about what constitutes or hinders effective incorporation of technology in classroom practice. The chapter also aims to examine the effectiveness of an innovative, in-service teacher training program in helping teachers to implement IT in teaching. The context for the study was the governmental imperative to integrate IT into ELT. In the first part of the paper, the context and the background to the in-service program is described. Qualitative and quantitative data are then presented which highlight the factors that both facilitate and impede curriculum innovation and change. In the final part of the chapter, the authors explore some of the ways in which the study can inform teacher educators and curriculum developers in providing in-service teacher education in IT to English language teachers.

Although this was a small-scale, preliminary study, it provides some useful insights into in-service teachers' beliefs about technology and its use in a more traditional transmission-oriented pedagogic culture. It shows that the English teachers were aware of the benefits of IT in raising the motivation of students and the quality of their learning, but they were also very concerned about the amount of time taken to produce IT-based materials. Because of a heavy workload and high teacher-student ratio, it was difficult for the teachers to find time to explore the possibilities, practice new skills, or try out new teaching ideas using IT. Furthermore, it was found that teachers were generally worried that they lacked the skills and knowledge to use IT effectively in teaching English, and they did not know where to find suitable computer-assisted language learning materials.

In their chapter, Nunan and Wong demonstrate the importance of understanding what IT and innovation mean for teachers, and of exploring teachers' beliefs and reactions to change. They also show how professional development is critical for helping teachers not just to develop a stronger sense of community but to see the pedagogic possibilities of IT. They point to the work of Cuban (1997), who argues that teacher educators and policy-makers need to "involve teachers deeply and continuously in on-site learning; hang in with them as the inevitable squalls of turmoil blow and recede; and finally, have patience, for such changes in belief and practice will take years" (p. xiv).

Similar concerns about the nature of teachers' professional development and about their needs are raised in the next chapter in this section, entitled "The Telenex Pedagogic Primary Grammar: Theoretical and Navigational Issues." In this chapter, Allan presents *TeleNex,* an innovative online resource designed to promote the professional development of teachers of English in Hong Kong. First established in its embryonic form in 1993 amid a climate of general concern over language standards, *TeleNex* comprises both a primary and secondary website and is available to all teachers in Hong Kong. The secondary website was established from 1993–98, and in the course of its five-year gestation, many lessons were learned about website architecture, navigation, interface design and the dynamics of computer mediated communication (CMC). These lessons informed the development of the primary website, which was developed from 1998–2002.

Allan begins his chapter by briefly outlining the main features of *TeleNex*. The main focus is on the theoretical and navigational issues associated with the development of one component in particular, a hypertext pedagogic grammar (*PrimeGram*), which was designed specifically for primary teachers. *PrimeGram* aims to provide English teachers at primary level with information about English grammar that will enhance their understanding of and proficiency in English and, help them teach more effectively. Therefore, the design criteria had to take into consideration the existing knowledge and background of the target users. Initial research indicated that Hong Kong primary teachers were inclined to see grammar as a set of (seemingly arbitrary) rules rather than as a set of resources for meaning making. The challenge was to help teachers develop a more coherent framework for thinking about grammar than a traditional/structural approach that emphasizes the forms and rules of grammar. Building on the work of Halliday (1985), *PrimeGram* took as its starting point the assumption that every time we use language, we do three things simultaneously: we talk or write about something, i.e., we use language to talk about our experience of the world around us; we talk or write to someone, i.e., we use language to interact with other people; and when talking or writing, we make links between the elements of language in order to create whole units of meaning, or coherent texts. From a pedagogical perspective, viewing language according to these three dimensions also provided a useful way for teachers to introduce elements of the language system to students.

On the technical side, the use of hypertext and multimedia posed a number of special problems, the requirement that navigation be simple, intuitive, and consistent proving to be one of the greatest challenges. The mechanisms that are currently in place have evolved through eight years of trialing and adaptation in response to users' feedback. The final section of the chapter touches on the question of sustaining the network from the dual perspectives of electronic database management and continued user involvement.

The final chapter in this section and in the book, entitled "The Standards Solution: The Role of Standards in Developing IT Professional Competencies and Communities of Practice in ELT," takes a critical look at various IT-related competency standards being introduced for English language teachers in pre-service programs and in education systems around the world, and the particular problems associated with their implementation at the chalkface. The chapter focuses on the different uses of IT standards in the ELT community — *setting* common expectations for professional practice, *raising* levels of professional knowledge and skill, and *meeting* agreed benchmarks for IT-based developments in the field. The author argues that if such IT standards are to be successfully realized, policy-makers and administrators need to budget for more than just the right hardware. They need to take into account the nature and the "lived' experience of technological innovation as perceived from the user's perspective. In innovation terms, IT is often seen by teachers as (yet another) change, imposed or exhorted from above rather than enhancement-inspired by experience or observation. Teacher stress, lack of collegiality and leadership, and the sheer scale and pace of change in IT also influence the effectiveness of the take-up of new technologies in the field. Stressed teachers do not have the energy to take risks, to try out new ideas, to accept the inevitable, if temporary incompetence accompanies technological experimentation. Systems do need to establish agreed expectations and standards for IT teacher competence, but more important still is structural support and a long practice time, the acknowledgment that professional competence is both experiential and theory-driven. The chapter concludes that many IT initiatives need to be better aligned with models of successful innovation and professional development and need to take more systematic account of teachers' (and students') actual experiences with technology and with innovation.

CONCLUSION

This edited collection of papers represents far more than just a collection of case studies on IT use in English language teaching. The volume provides an exploration and critique of many of the most fundamental assumptions in the ELT field about IT as a tool for learning and teaching, and for professional enhancement. As the title suggests, it is actually much more concerned with IT as innovation than with specific IT practices as such. I hope the reader finds that the actual practices are interesting exemplars of IT-based developments in English language teaching around the world, but the broad and long-lasting appeal of the book is in its concerns with understanding the strengths and limitations of IT as innovation, as pedagogy and as professional development.

NOTE

1. According to the Australian Council for Computers in Education the convergence of communications and computer technology in the IT industry means there is no longer any sense in distinguishing stand-alone artefacts from connected ones.

SECTION II

Information Technology as Innovation

2 New Frontiers in Technology and Teaching

Denise E. Murray

Making paper copies of anything is a primitive use of such machines (computers) and violates their very spirit. (Toffler 1980, 205)

I think there is a world market for maybe five computers. (Thomas J. Watson, founder and chairman of IBM, 1943, quoted in Bell 1997)

The computer will take over everywhere from the operation of thinking, leaving the brain to lie fallow, as the mechanistic technologies of the nineteenth century have already done with the body. People are becoming increasingly zombie-like. It looks as if their brains have already been removed and they are merely functioning on their spinal cords. (Jean Baudrillard 1990, 213)

Intuition, insight, and learning are no longer exclusive possession of human beings: any large high-speed computer can be programmed to exhibit them also. (Simon and Newell 1958, 6)

We will "connect every classroom to the Internet by the year 2000 and help students become technologically literate." (President Clinton, 1997 State of the Union Address)

INTRODUCTION

These predictions, which today seem ludicrous, were based in understandings of the nature and potential of computers that were current at the time. Inaccurate predictions concerning new technologies have not been unique to the use of digital technology. Inventors of earlier communications technologies, constrained by their own visions for their invention, were equally unable to predict how human beings would adopt a technology, amplify its use, and incorporate it into social life. Alexander Graham Bell, for instance, believed he had invented a technology for business and a broadcasting device; Samuel Morse believed he had invented a technology that would help people communicate with each other. Early use of the radio for military communication led to point-to-point communications linking amateur enthusiasts; later it developed into a commercial broadcast system. The telephone, in contrast, became an instrument for social interaction. So, Alexander Graham Bell, who invented the telephone, thought he had invented the wireless, and Samuel Morse, who had invented the wireless, thought he had invented the telephone.

The inventors of the computer, such as Alan Turing, were looking for ways to mechanize mathematical calculations, hence the name of the machine. Very quickly, however, human users discovered that they could use dumb terminals logged onto the same mainframe computer to send messages to each other, that is, to communicate. The precursors of current networking systems were originally designed to send data: ARPA was designed to connect researchers working on US Department of Defense contracts so they could share data; IBM's VM network was designed so computer programmers working on the same project could send data files to each other. In each case, users quickly added messages and very quickly, the networks became used for human-human communication, not just the transfer of data. Either the inventors themselves, or users during the popularization of a communications technology, develop a simplified view of the social forces surrounding the new technology and thus are unlikely to predict future uses of the technology. Nunberg (1996) argues that most predictions take the most rapid point of change and extrapolate from there to the future, without noticing other forces that may be regrouping. In fact, "technologies are often developed in response to the agendas of powerful social actors. Initially, they shape themselves to the contours of custom; ultimately, they follow paths selected through struggles among groups seeking to turn technologies to their own interests" (DiMaggio, Hargattai, Neuman, and Robinson 2001, 327).

THE NATURE OF THE MEDIUM

When we talk about using new technologies in the language classroom, what technologies are we referring to? Within and outside education, a variety of terms are used to describe the new phenomena. Before we can begin to understand how we might shape and be shaped by these new technologies, we need to deconstruct the plethora of terms, knowing as we do that the way we name a concept will influence the way we perceive it. The metaphors we choose almost always come from previous technologies, and so we find ourselves hobbled to the past while we try to imagine the future; and as we know from other settings, metaphors shape our understanding and view of the world (Lakoff and Johnson 1980; Lawler 1995).

If we use, for example, the term *email*, are we restricting ourselves to thinking about this type of communication as a form of letter? If we use the *desktop* as a metaphor, are we restricting ourselves to a two-dimensional desk, knowing that many people "file" information in piles on their desk and even on the floor? Is the fact that we use *folders* to name a collection of *files* restricting us to thinking about filing cabinets? While these may have been useful metaphors in the early days of computer technology, do these office metaphors serve us well today, when we are living in a hypertext world? How can we represent the relationship of the personal information we store (such as email, documents) if we are still locked into using metaphors from a paper world? Often, the metaphor of a tree is used to explain hypertext and indeed, the word branch(ing) is the term used to describe hypertext paths, solidifying that metaphor. Does this metaphor help, or hinder, both our understanding of hypertext and our ability to develop new ways of using the technology? The place at which we "find" a new branch we want to go to is described as a link; yet link is a different metaphor from trees/branches, conjuring up as it does chains and fences, a more two-dimensional, rigid world than that of trees. The tracing of paths has also been referred to as threads, a metaphor that conjures up the thread one uses to find one's way out of a maze. The effect of metaphor on our understanding of the new technologies is as yet an unexplored area (but see Murray, 2004, for a brief exploration). However, despite the limits our metaphors may impose on our understanding and creativity, new tools are being introduced that try to explode (or at least make transparent) some of the metaphors. For example, in discussion lists, we now have threaded discussions that do represent visually the relationship between posted messages — as long as the message poster links to the correct thread. One of the ironies of metaphor use is the example of "wired", which has been used in phrases such as "the wired generation", and is the name of the leading-edge computer magazine *Wired*. Despite the limitations of this metaphor to a physically connected world, researchers and developers have been able to go beyond the metaphor, take up the wireless mobile phone and begin conceptualizing and developing wireless computer technology (see below).

The overarching term for the new digital technologies is either *information technology* (IT) or *information and communication technology* (ICT), the latter now taking the place of IT since email became the *killer app* (the most used application). Similarly, the 1990s have been referred to as *the information age*, and newspapers decry our *information overload*. National leaders, however, invoke the notion of the *knowledge nation*. Castells (1998) has shown that the developed world is moving from the industrial age to the information age, when economic growth is based on the exchange of information and the development of knowledge. However, this focus on information needs to be deconstructed. As referred to above, the early technologies focused on file transfer of data. *Data* refers to the representation of something assumed to be fact; however, the fact is not necessarily accurate. Data becomes information only if it has potential value and the potential to impact upon an individual's knowledge. As we interact with information and actively reconstruct it to become part of what we already know, we create knowledge that is structured and hierarchical, and socio-culturally grounded. As we develop expertise and experience with certain sets of knowledge, we develop organic wisdom. Agre (1995) explains why the term *information* continues to be used to subsume data, knowledge, and even wisdom as follows:

> The term 'information' rarely evokes the deep and troubling questions of epistemology that are usually associated with terms like 'knowledge' and 'belief.' The concept of information, then, carries a certain connotation of neutrality—it is homogenous and not controversial. The reality, of course, is more complicated.

When we use IT, or even ICT, we invoke only a potential for knowledge and wisdom. However, if we carefully examine the use of such terms in the popular (and even educational) literature, we find that information is used synonymously for data and even for knowledge. A web search illustrates this For example, a search on the word *information* retrieves

URLs related to information technology, information overload, government information and search engines. One news item (EContent 2003) is entitled "City of Chicago Deploys FAST Data Search on Public Information Site." The article switches between the use of the terms *data* and *information* almost randomly. A closer analysis reveals that the term *data* is used in the name of their search engine and for monitoring or filtering data; however, once the data have been retrieved, they use the term *information*, even though it has only been retrieved by a computer program (in the same way the computer program monitored and filtered data) and has not demonstrated potential value to a human actor. Of course, the term *IT* was originally intended in a technical sense; as in information theory, in which information can refer to the amount of information in a TV signal or the number of bits and bytes. However, it has been co-opted for more general use, leading to the confusion among the terms *data*, *information*, *knowledge*, and *wisdom*.

In the field of language teaching, the earliest term used in this area was *computer-assisted language learning* (CALL), which clearly places the computer at the same level as other audio-visual equipment, as a tool to assist the teacher. However, as Warschauer (2001) points out, the computer is not merely a tool for learning English; English for many learners is a tool for learning and using the new technologies. By using the word *computer*, we foreground a particular device, one that may eventually be quite transparent to the learner (see later discussion about convergence). Similarly, the term *multimedia* tries to capture the notion that audio and video are being employed for teaching/learning, but in doing so foregrounds the media as merely a tool. However, this again traps us in the language of old technologies.

This review of past and present conceptualizations in ICT and language teaching raises the question of what awaits us in the future. The next two sections address this issue in relation to advances in technology and advances in institutionalizing technology. The first deals with the development of new technologies; the latter refers to the ways current technologies are being shaped by social forces.

ADVANCES IN TECHNOLOGY

As mentioned earlier, it is impossible to predict the way technologies will be shaped by socio-economic, political, and cultural human agency. We do, however, have some notion of the technologies that are currently in the developmental stages or being used by early adopters, early adopters being the small percentage (around sixteen percent) who take up an innovation because they are captivated by it and its potential (Rogers 1995). The developments I will refer to are ones that have the potential for use in language education and so are likely to be adopted by teachers.

Broadband

One trend we are already seeing to some degree is the use of broadband that allows for visual and moving data to be transmitted at speeds that are sufficiently fast for people to feel comfortable. However, the rollout of broadband has not been without difficulties. In addition to its cost, there have been documented failures in delivery. However, the potential for more flexible, multimedia delivery of language instruction will rely on broadband. Without it, we can't have streaming video, and even audio files and highly intensive graphical interfaces take too long to download. Yet, language education relies on realia, and one of the advantages of the new media is the ability to have realia "just-in-time," that is, when the learner and teacher need it.

Convergence

How many devices do you have that you can use for entertainment or communication? For many of us, digital technologies have complicated our lives because of the overwhelming number of new devices that need shelf and table space in our homes and offices: we have a fax machine for sending copies of print materials; we have a computer to surf the web and send email; we have a mobile phone to receive and send voice (and print) messages while traveling; we have a TV for receiving free-to-air and pay-per-view entertainment; we have a CD or DVD player to listen to our favorite music; our children have a game machine for playing computer games, and so the list goes on. According to *Scientific American* (2000), help may be closer to hand that we think. *Convergence*, the integration of audio, video, and data communication into a single source that can be received on the same device via the same connection, has been predicted for decades. The first such convergent device was around in 1939, when television was formally demonstrated at the New York World's Fair. The device used was actually a combination television-radio-recorder-playback-facsimile-projector, a device well ahead of its time and never sold commercially. The desire, at least on the part of manufacturers, is to provide movies, TV, music, and communication on our home big screens, our computers, our wristwatches, and our mobile phones. In other words, the goal is to provide any entertainment and communication anywhere, anytime, and on any device. This desire to link content with delivery mechanism has lead to America On Line (AOL), an Internet deliverer, merging with Time Warner, a conglomerate that controls news and entertainment, that is, content. This will, of course, lead to a situation in which those who create and control the content also control its delivery.

Wireless communication

One of the constraints of current computer technologies for email, net surfing, word processing, and so on, is that, even with laptops, batteries have a short life, and to be connected to a network, one needs either a direct connection or a connection over a telephone line. This has restricted use in countries where electrical power and telephones are either scarce or unreliable and, even in countries where they are readily available and reliable, it means anytime, anywhere communication is still not possible. The use of wireless technology (that is, radio as opposed to telephone) has been available for several years; however, it is only in very limited areas, such as Silicon Valley in California, where both the density of population and the prolific use of online work warrant the installation of wireless transmitters on light poles. Like cell (mobile) phone technology, the transmissions are over short distances, and so many receiver/transmitters need to be installed at great cost. However, we have seen the tremendous, rapid growth of cellular phones, especially in countries such as China where there are few landlines. Therefore, it is quite likely that wireless technology for online computing will also be attractive in such countries. Such technology will allow learners to bring their computers to the classroom and log on from any room. Teacher and learner will be connected, without the necessity for stationary workstations and a multitude of wiring. However, to cut costs, others are predicting that the computer of the future will be a "dumb" terminal, connected to a network that will store all the documents and programs. Such terminals would be considerably cheaper and therefore provide greater access to ICT for the poor. We lose flexibility as we increase complexity (in this case access to many modes of the technology), as discussed later in the section dealing with the issue of complexification.

Miniaturization

A complementary technology development is the handheld computer that now can be used for almost all the functions of a desktop computer. SRI International, an independent nonprofit research and development organization involved in researching and creating innovative technology and business solutions, has joined with Palm to investigate the use of wireless handheld computers for learning in US schools. Based on their research projects, they have found that these devices augment physical space, leverage topological space, and create "a more distributed systems peer-to-peer network topology" (Roschelle and Pea, in press).

A new development that combines miniaturization, wireless communication and convergence is the grid developed in 2003 at CERN, the European particle physic laboratory where the World Wide Web was born. The grid, like electrical grids, will provide online connectivity on demand. Using a mobile phone for wireless access and a handheld for display (or even the new generation mobile phones for display), we will be able to get data, read and send email and so on, without the need for a powerful PC.

Discussion support software

There have been several attempts at using computers to organize discussions, but most have floundered because humans have preferred to interact with as little technical mediation as possible. Some recent systems in current research and development may hold out more hope for such software. IBM is experimenting with two systems. One, called *Babble*, displays a window in which dots within a circle show who is in the current conversation. The closeness of the dots to the center of the circle indicates how recently each person has contributed to the discussion. There is another window where participants can nominate and prioritize discussion topics. A transcript from the discussion indicates both what was "said" and the time lapses between contributions. The second, more recently developed system expands on *Babble* and includes the ability to post "Post-it" notes and provides information about the physical location of participants (IBM Social Computing Group Website n.d.).

INSTITUTIONALIZING OF TECHNOLOGIES

The social impact of the Internet and other ICT technologies depends on the ways society shapes and moulds these technologies. While many aspects of ICT are still in their early stages of development and, historically, are rather like the point-to-point stage of radio communication, we already have evidence of social (and political) structuring of their uses.

I will deal with only a few issues in the institutionalization of technology, ones that already do or are likely to impact on language education: borderless community, access, complexification, and computer literacies.

A borderless community

By its very design, the Internet is borderless, and certainly, the earliest denizens of the net likened it to the opening of the West in the United States — open plains, the ability to take up land anywhere, and minimal law and order. However, just as the pioneers ignored the original inhabitants and their rights, early Internet users ignored and still fight against censorship and restrictions based on the socio-cultural values of different communities. We have already seen individual countries filtering sites they consider to be inappropriate politically or culturally. In Germany, an Internet Service Provider is required to block child pornography sites from German subscribers. So, in effect, we have national borders being established within the Internet.

Access

Access involves a number of issues, all of which are of interest to language educators: issues of physical access, the language medium, and navigation skills:

> Access to information technology is not universal; a minority of the world's population has access to the Internet and even within this minority, some cannot receive information in a form (for example, language) that is accessible to them. (Murray, 1999, 5)

Access is usually discussed as physical access to computer technology and is often referred to as the digital divide. Data indicate that physical access to the technology is divided along socio-cultural and economic lines, the poor, minorities and the less-educated in technologically advanced countries having less or no access. Internationally, the divide follows the division of wealthy economies (Murray 1999). Thus, for example, there is minimal access in Africa but far greater access in most of Europe and North America. Although access to ICT is increasing each year, still less than ten percent of the world's population is online. What we notice from recent trends, however, is that the increases are most remarkable in the less technologically developed nations. Growth in Europe, the US, Canada, Japan, and Australia, for example, is not as rapid as growth in China or India.

Because of the original dominance of ICT by the English-speaking world, especially the United States, English has been the dominant language on the Internet. Global-reach, an international online marketing company, collects and organizes data by language to facilitate global marketing by corporations. Their data are based on the assumption that people speaking the same language but living in different countries will form their own language communities. Therefore, these data do not reveal whether an email from a Japanese person to a Japanese colleague was written in English or in Japanese. There is anecdotal information that many people use English even when they share another common language. However, language use patterns among bi- and multi-linguals are complex. Consequently, the statistics in Figure 2.1 can only be regarded as approximations. Despite these reservations, the data indicate that English is the language of the Internet. So, for some time, English has been the passport to going online, especially to gathering information on the World Wide Web.

However, these data are of particular interest when we compare them with data over the last few years and see the growth in non-English speaking Internet use (see Table 2.1).

Furthermore, from July 2002 to July 2003, English use decreased from 40.2% to 35.2%, while Chinese increased from 9.8% to 11.9%. Chinese is the fastest growing language in use on the Internet, as can be seen from Table 2.2. The data in the following table come from a number of sources and have been presented in tabular format for ease of comparison.

So, while both Internet users and webpages have increased across all language groups, the position of English has actually decreased as a percentage of the total Internet users and webpages.

While physical access and linguistic access are clearly important issues for language educators to consider, yet another form of access impinges

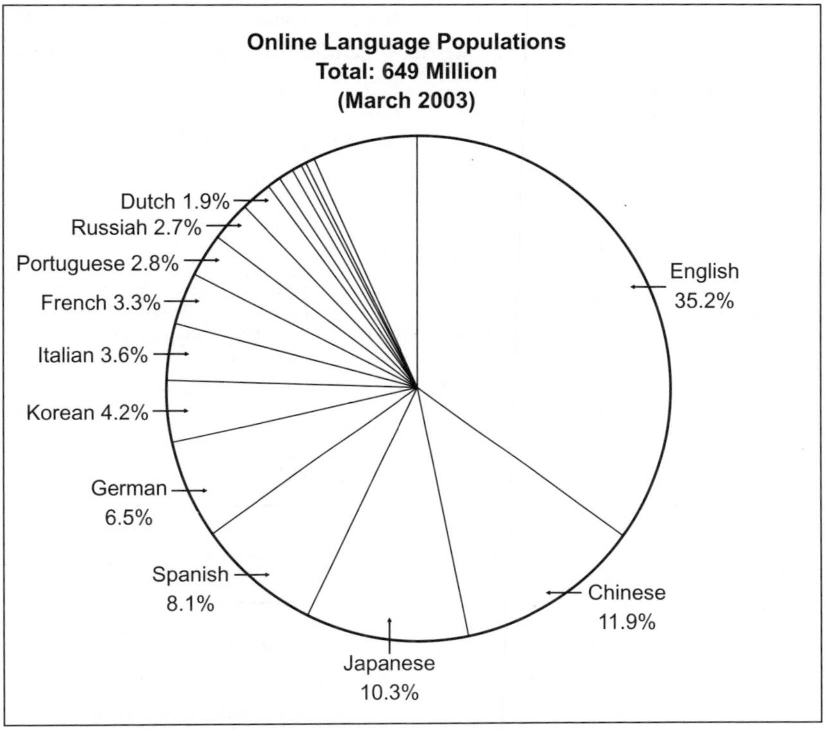

Figure 2.1 Online populations by language

Note: From *Global Reach (2003). Internet Statistics (by language)*, retrieved July 2, 2003, from http://global-reach.biz/globstats/index.php3

Table 2.1 Internet users over time

Language	1997 Network Wizards 1997	2001 Internet Software Consortium 2001	2002 Global Reach 2002	2004 Global Reach 2004
English	128 million	220.4 million	225.5 million	287.5 million
Non-English	88 million	292.7 million	335.5 million	516.7 million

Note: 1. The data in column 4 are from *Global Reach (2002). Internet Statistics (by language)*. Retrieved July 9, 2002, from http://global-reach.biz/globstats/index.php3.

2. The data in column 2 are from Network Wizards (1997). Internet domain survey. Retrieved January 10, 2002, from http://www.nw.com/zone/www.top.html.

3. The data in column 3 are from Internet Software Consortium. (2001). Internet domain survey. Retrieved January 10, 2002, from http://www.isc.org/ds/
4. The data in column 5 are from *Global Reach (2004). Internet Statistics (by language)*. Retrieved August 27, 2004 from http://global-reach.biz/globstats/index.php3

Table 2.2 Webpage top five languages

Language	1997 Network Wizards 1997[1]	2000 Vilaweb.com 2000
English	83%	68.4%
German	4%	5.8%
Japanese	1.6%	5.9%
Spanish	1.1%	2.4%
Chinese	<0.1%	3.9%

Note: The data in column two are from Network Wizards. (1997). Internet domain survey. Retrieved January 10, 2002, from http://www.nw.com/zone/www.top.html. Copyright 2002 by Internet Software Consortium. The data in column three are from *Language Study*, by Vilaweb.com. 2000. Retrieved January 10, 2002, from http://cyberatlas.internet.com/big_picture/demographics/print/0,,5901_408521,00.html.

on our work. Research has shown there is also considerable variance in abilities to locate content on the web (e.g., Hargittai 2002; Murray 2003). Hargittai (2002) argues that "merely offering people a network-connected machine will not ensure that they can use the medium to meet their needs because they may not be able to maximally take advantage of all that the Web has to offer … Although providing Internet access may help alleviate some problems of the digital divide … a second-level digital divide exists relative to specific abilities to effectively use the medium" (pp. 13, 14). In some settings, such as adult ESL programs or teaching in developing countries, teachers cannot assume that their learners are proficient computer users or web surfers. Language educators in such settings are therefore in the position of being able to introduce this new technology to learners, not just as a tool in language learning, but as a tool and language they will need for their future lives. Research in the Adult Migrant English Program in Australia, for example, has shown that these learners are empowered by (but also often initially intimidated by) learning to use email. They quickly learn they can use it with their teachers and even more importantly, to connect with family and friends in their home countries (Murray and McPherson, 2003).

Complexification

Karl Ernst von Baer, who discovered the mammalian egg cell in 1827 and wrote the founding treatise of modern embryology in 1828, developed a law that is now named after him (Gould 2001). Von Baer's law epitomizes the central issue in current debates over the openness of the Internet, both in its accessibility and in its usability. The very structure of material reality imposes a principle of trade-offs in both nature and human affairs: One always gives something in order to gain. In particular, we usually pay for complexity by surrendering flexibility. One such example is word processors. Early programs were not very complex, but users could control many of the features they wanted in their documents, such as lists and headings. As word processors have become more complex, with more and more features, the ability to control certain features has become increasingly difficult. The programmers have decided how I should write a letter or how I should write lists, and the program automatically inserts features, using hidden codes. It often takes considerable expertise with a word processor to be able to override some of these "features." When the web was small and less complex than it is now, finding information on a topic was relatively simple with a search engine such as Alta Vista or Lycos. The search usually resulted in a short list of URLs that had the information one wanted. As the web has become more and more complex, such search engines are usually not powerful enough and one needs to use a search engine that searches other search engines, such as Google. In addition, the search terms need to be more and more precise to avoid receiving thousands of URLs. As the Internet has become more complex, its flexibility as a tool that can be used as "anything, by anyone, at anytime" has been reduced. Increasingly, users need more complex knowledge and skills to use the Internet.

Computer literacies

ICT has resulted in new text types, and new structures and with visual features increasingly becoming used to convey information, not just to support printed text, a situation that has been described as "a tectonic shift" (Kress 1997) and "the breakout of the visual" (Bolter 1996). According to Kress (2000, 337), it "is now impossible to make sense of texts, even of their linguistic parts alone, without having a clear idea of what these other features [audio, video, images] might be contributing to the meaning of a text." To help learners navigate, comprehend, and even construct these new text types will require, among other characteristics of texts, a clearer understanding of the grammar of visual literacy. First we will need linguistic descriptions of these new text types and then strategies for teaching these new text types, just as we have a variety of teaching approaches to the teaching of written texts (e.g., text-based approaches, rhetoric, process writing).

IMPLICATIONS FOR LANGUAGE EDUCATION: INSTRUCTION

Possible changes to instruction resulting from the above potential developments can be categorized by language to be taught, organization of instruction, materials design, and the future of ELT.

Language to be taught

As a result of the new text types with which learners need to be familiar both inside and outside the classroom, at a minimum, learners will need to become proficient in a variety of new literacies.

They will need to read webpages. Webpages consist of a variety of text types, such as advertising, pop-up windows, menus, flashing windows, visuals, and even audio. While many researchers and organizations are working on design principles for webpages (e.g. International Visual Literacy Association), we cannot consider webpages to have yet developed a specific *genre*, that is, culturally typical ways of engaging with recurring situations that are learned through social contexts (Martin 1993). Learners are confused by these less well-known text types, and teachers have no descriptions on which to base their teaching (Murray 2003). In addition to being able to read webpages, learners must search the web for information, which requires that they learn not only how to use search engines but also how to evaluate the results of such a search. They also need to learn how to write online: filling out forms, writing their own webpages, and word processing and desktopping their reports and other work. As more learners go online, however, what we may be teaching is not the generic skills but the contrastive rhetorical systems that differentiate different literacies in different cultural groups. There is currently no research on such contrastive rhetoric, but, since genres are culturally constructed, it is likely they will develop differently across cultures.

In addition, they will need to be taught the conventions of email, conventions that are still developing.

Organization of instruction

With wireless devices, whether handheld or laptop size, teachers will be able to integrate ICT into their classroom just as they currently integrate the OHP or the whiteboard and chalk. One of the major constraints in using current technologies is that they are not ready-to-hand and available for "just-in-time" learning. Time in the computer lab has to be planned for and inevitably is not available just at the moment the teacher wants to use a webpage or email in her instruction. If the discussion software proves to be supportive of language learning, teachers could use it for online learning, a learning that would include audio and video via broadband networks.

Materials design

If learners have difficulty with certain text types and navigations, then it is incumbent on materials designers to develop ICT language learning materials that students can comfortably use. If CALL writers are writing for anywhere, anytime handheld wireless devices, they will need to consider the difficulties of miniaturization, especially for language learners not familiar with either the language or the alphabetic script. With the advent of broadband, we can expect materials that are truly realia. Designers will need to design for the different devices — materials that work optimally on a broadband network with high-speed cable to a fixed line will not be optimal for a handheld wireless device. As Van Baer notes, as we become complex we lose flexibility and vice versa.

THE FUTURE OF ELT

As noted previously, although English is still the language of global communication, other language groups are increasingly coming online.

This, in the initial stages, will mean a greater market for online delivery of courses including English language courses. However, over time, if Graddol (1999) is correct in his predictive statistics, by the middle of the twenty-first century, English may no longer be the dominant language. While it is, we should take this opportunity to harness both the technology and the language to help our learners. Because of projects such as those conducted by Cummins and Sayers (1995) or advocated by Warschauer (2001), we can use the technology and English to help break down the barriers between communities, to help learners engage in critical, collaborative, meaning-making projects. However, we must still be mindful that barely ten percent of the world's population is online, and the more we make English and technological literacy requirements for successful employment, the more we exclude those with access to neither. If we take a critical stance, though, perhaps our learners will also reread and rewrite the world to foster diversity and inclusion.

CONCLUSION: TEACHER'S WORK

Teachers have reacted to the introduction of ICT in education in a number of ways: fear of becoming redundant, resistance to having to learn new ways of teaching, concern about learning the technology, uncertainty about whether the technology will dominate pedagogy, or enthusiasm at the potential of the new technologies to facilitate learner autonomy. While the fears and concerns are predictable, given the research literature on change and innovation (see, e.g., Fullan and Stiegelbauer 1991), ICT introduces additional complexities into teacher response to change. Because ICT is still not an anytime, anywhere technology, teachers have to plan in advance to use the technology, often having to book a computer lab weeks in advance. This constrains the immediate use of new materials in response to unexpected classroom events, responses that are possible with other pedagogical tools and realia. For example, if a particular topic arises in a class, a teacher can find the information or object ready for the next class session; however, with ICT, the teacher may have to wait until her scheduled time in the computer lab. By then, the class may have moved on to another task. Some earlier technologies (e.g., video) required some understanding of technology; however, ICT requires teachers have far more technical knowledge to be able to use it effectively. Additionally, it is still a rather unstable technology, small errors resulting in system failures. Many advocates of ICT in language education cite its fostering of learner autonomy as an advantage of CALL over traditional classroom instruction. While studies have shown that learners can become independent learners (see, for example, Jones, Valdez, Nowakowski, and Rasmussen 1995), others have shown that learners need computer work to be scaffolded (see, for example, Murray 2003). Because language learning occurs through interaction and in classroom-based instruction, such interaction is between teacher and student, student and student, or student and text, and the quality of classroom interaction has an effect on language learning. The teacher's role is one of providing learners with opportunities to engage with others — people and texts. One way of conceptualizing the teaching-learning situation is as a triangle of teacher, learner, and materials, embedded in a context (see Figure 2.2).

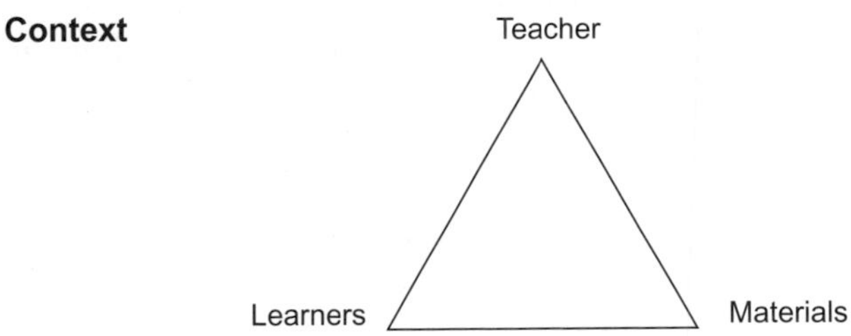

Figure 2.2 The teaching-learning context

The question of interest is whether the major interaction in the classroom will be directly between teacher and learner and among learners, or whether this interaction will be mediated via the materials, in this case, computer-based materials. Even when teachers are explicitly assisting learners to negotiate with texts, teacher and learner are central. With print materials, we have seen that some classrooms are mediated via materials, as when, for example, the teacher works entirely from a textbook or has students complete worksheets. Although novice teachers (and less engaged teachers) may have resorted to such strategies with print-based materials, the temptation to do so with computer-based materials is much greater. Firstly, much of the literature on CALL claims one of the benefits of CALL is student autonomy. Teachers could then quite easily respond by mediating language interaction through the computer. Secondly, CALL programs and computer-mediated communication (CMC) can give the appearance of providing learners with opportunities for interaction. But, this interaction, unless carefully scaffolded by the teacher, can have the appearance of interaction, while not providing learners with authentic opportunities for negotiating meaning.

While teachers' work will undoubtedly change with the use of ICTs in the classroom, as it did with the introduction of paper, print, and photocopying, the essential humanistic elements of classroom interaction are unlikely to change. Students and teachers will continue to develop relationships, upon which learning builds; teachers will continue to need to scaffold learning. As with earlier technologies, computers can support authoritarian classrooms just as easily as they can support constructivist classrooms. But, as Cummins and Sayers (1995) have shown, technology, harnessed for transformative pedagogy, can create intercultural global learning networks that can "challenge coercive social forces" and "develop students' academic language and critical literacy" (Cummins 2000, 546). The work of teachers will remain one of adapting materials for classroom use, of selecting materials that suit learners' needs, and providing an environment that supports learning. The challenge lies in focusing on pedagogy, not on technology. As Warschauer (2001, 8) concludes:

> We will fulfill the best use of computers in the classroom when we allow and encourage students to perform the most real tasks possible, to take advantage of the power of modern information and communication technologies to help try to change the world in ways that suit students' own critical values and the interests of humankind.

NOTE

1. Network Wizards was bought out by Internet Software Consortium.

REFERENCES

Agre, P. (1995). Institutional circuitry: Thinking about the forms and uses of information. *Information Technology and Libraries, 14*(4), 1-6.

Baudrillard, J. (1990). *Cool memories* (C. Turner, Trans.). London: Verso.

Bell, G. (1997). Laws of predictions. Retrieved Jul 3, 2003, from http://research.microsoft.com/acm97/gb/sld001.htm.

Bolter, J. D. (1996). Ekphrasis, virtual reality, and the future of writing. In G. Nunberg (ed.) *The future of the book* (pp. 253–72). Berkeley, CA: University of California Press.

Castells, M. (1998). *End of millennium*. London: Blackwell.

Clinton, W. J. (1997). *State of the Union address*. Retrieved July 3, 1999, from http://www.whitehouse.gov/WH/SOU97/

Cummins, J. (2000). Academic language learning, transformative pedagogy and information technology: Towards a critical balance. *TESOL Quarterly*, *34* (3), 537–48.

Cummins, J. and Sayers, D. (1995). *Brave new schools: Challenging cultural literacy through global learning networks*. New York, NY: St. Martin's Press.

DiMaggio, P., Hargattai, E., Neuman, W. R., and Robinson, J. P. (2001). Social implications of the Internet. *Annual Review of Sociology*, *2*, 307–36.

EContent. (2003). City of Chicago deploys FAST Data Search on public information site. Retrieved July 2, 2003, from http://www.econtentmag.com/Articles/ArticleReader.aspx?ArticleID=4628

Fullan, M. G. and Stiegelbauer, S. (1991). *The new meaning of educational change*. 2nd. ed. London: Cassell.

Global-reach (2003). *Global Internet statistics (by language)*. Retrieved July 2, 2003, from http://global-reach.biz/globstats/index.php3.

Global-reach (2002). *Global Internet statistics (by language)*. Retrieved July 9, 2002, from http://global-reach.biz/globstats/index.php3.

Global-reach (2004). *Global Internet statistics (by language)*. Retrieved August 27, 2004, from http://global-reach.biz/globstats/index.php3.

Gould, S. J. (August 27, 2001). *What only the embryo knows*. Retrieved August 27, 2001, from http://www.nytimes.com/2001/08/27/opinion/27GOUL.html.

Graddol, D. (1999). The decline of the native speaker. In D. Graddol and U. H. Meinhof (eds.) *English in a changing world (AILA Review 13)*. Biddles, Ltd.

Hargittai, E. (2002). Second-level digital divide: Differences in people's online skills. Retrieved July 9, 2002, from http://firstmonday.org/issues/issue7_4/hargittai.

IBM Social Computing Group Website (n.d.). Retrieved May 4, 2002, from http://www.research.ibm./com/SocialComputing/

Internet Software Consortium. (2001). Internet domain survey. Retrieved January 10, 2002, from http://www.isc.org/ds/

Jones, B., Valdez, G., Nowakowski, J., and Rasmussen, C. (1995). New times demand new ways of learning. In *Plugging in: Choosing and using educational technology*. North Central Regional Education Laboratory. Retrieved July 9, 2002, from http://www.ncrel.org/sdrs/edtalk/toc.htm

Kress, G. (1997). Visual and verbal modes of representation in electronically mediated communication: The potentials of new forms if text. In I. Snyder (ed.) *Page to screen: Taking literacy into the electronic era* (pp. 53–79). London: Routledge.

Kress, G. (2000). Multimodality: Challenges to thinking about language. *TESOL Quarterly*, *34*, 337–40.

Lakoff, George and Mark Johnson. (1980). *Metaphors we live by*. Chicago: University of Chicago Press.

Lawler, John M. (1995). Metaphors we compute by. A lecture delivered to staff of the Informational Technology Division, University of Michigan. Retrieved July 2, 3003, from http://www-personal.umich.edu/~jlawler/meta4compute.html.

Martin, J. (1993). Genre and literacy: Modelling context in educational linguistics. *Annual Review of Applied Linguistics*, *13*, 141–72.

McPherson, P. and Murray, D.E. (2003). *Communicating on the Net*. Sydney: NCELTR Publications.

Murray, D. E. (2004). The language of cyberspace. In E. Finegan and J. Rickford (eds.) *Language in the USA: Themes for the twenty-first century* (pp. 463–479). New York: Cambridge University Press.

Murray, D. E. (2003). Materials for new technologies: Learning from research and practice. In W. A. Renandya (ed.) *Methodology and materials design in language teaching* (pp. 30–43). Singapore: SEAMEO Regional Language Centre.

Murray, D.E. (1999). Access to information technology: Considerations for language educators. *Prospect 14*(1), 4–12.

Network Wizards. (1997). Internet domain survey. Retrieved January 10, 2002, from http://www.nw.com/zone/www.top.html.

Nunberg, G. (1996). Farewell to the information age. In G. Nunberg (ed.) *The future of the book* (pp. 103–38). Berkeley, CA: University of California Press.

Rogers, E. M. (1995). *Diffusion of innovations*. New York: The Free Press.

Roschelle, J. and Pea, R. (in press). A walk on the WILD side: How wireless handhelds may change computer-supported collaborative learning. *International Journal of Cognition and Technology*.

Scientific American. (November 2000).

Simon, H. A. and Newell, A. (1958). Heuristic problem solving: The next advance in operations research. *Operations Research*, *6*(6).

Snyder, I. (1997). *Page to screen: Taking literacy into the electronic era*. London: Routledge.

Toffler, A. (1980). *The third wave*. London: William Collins.

Vilaweb.com. (2000). Language study. Retrieved January 10, 2002, from http://cyberatlas.internet.com/big_picture/demographics/print/0,,5901_408521,00.html.

Warschauer, M. (2001). The death of cyberspace and the rebirth of CALL. In P. Brett (ed.) *Call in the 21st century*. CD-ROM. Whitstable, Kent: IATEFL.

3

Making Curriculum Innovation Work: The Role of Information Technology Centers

Numa Markee

INTRODUCTION

As the World Wide Web and other information technologies (IT) become ever more widespread and pervasive in people's everyday lives, there is increasing pressure on educational institutions to use these technologies as resources for pedagogy. IT is undoubtedly enormously seductive; however, its use in educational contexts must be carefully thought out, particularly in the context of second language education. Issues of implementation include questions relating to the general function of IT in the context of curricular innovation, and what IT is good for and not so good for in ELT. In addition, it needs to be appreciated that large numbers of people playing a variety of quite different social roles are involved in making the use of IT a success and that end users will ultimately determine whether an IT-based project will succeed or fail. This means that issues of user support must be carefully planned and that adequate resources must be allocated to the implementation of a project over the long term.

Finally, and this is undoubtedly the most important point, we need to be able to justify why we think the use of IT is going to be advantageous. Given the complexity of the technology and its high cost, we should have a very clear idea of what we are letting ourselves in for when we decide to use IT. That is, we should be able to articulate why we think the benefits of using IT outweigh the substantial risks and, indeed, the financial and human costs associated with stepping into this technological brave new world.

This chapter therefore begins with a brief outline of a theoretical framework for understanding the process of curricular innovation and, more specifically, the role and function of IT in this process. This framework is illustrated throughout by drawing on the experience gained by the Language Learning Laboratory (LLL) at the University of Illinois at Urbana-Champaign (UIUC) in supporting IT-based curricular innovation in several foreign language departments. In addition, two examples of innovative programs designed to encourage end users to use IT are described in some detail. In conclusion, the larger implications of such initiatives are also discussed.

CURRICULUM INNOVATION: A THEORETICAL FRAMEWORK

We can try to understand how language education changes by asking the following interrelated questions: "Who adopts what, where, when, why, and how?" The answers to each constituent element of this composite question develop the architecture of an integrated, interdisciplinary model of curriculum innovation in second and foreign language teaching (Markee 1993, 1994, 1996, 1997a; see also Cooper 1989, from whom this framework derives).[1] For present purposes, the issues that are raised by each sub-question are briefly described, and then three of these sub-questions are discussed in more detail: What, Who and Why.

Who-type questions seek to identify the individuals who have a stake or interest in innovation X. Furthermore, these questions attempt to clarify the different social roles these people play out as innovation X is adopted. Questions about adoption focus on illuminating the decision-making processes that potential adopters go through as they decide whether to adopt or reject innovation X. Questions about *what*-related issues bring out what types of innovations are potentially being diffused. *Where*-questions focus not so much on geographical context as on socio-cultural context (Cooper 1989), for example, the kinds of cultural, political, administrative, educational, and institutional variables (among others) that may potentially constrain classroom innovation (Kennedy 1999). *When*-type questions address various issues related to the amount of time it takes for people to adopt innovation X. Answers to *why*-questions give details about participants' psychological profiles and provide information about the general characteristics of innovations that either encourage or inhibit their adoption. And *how*-type questions get at the managerial techniques that are used to implement educational change. Insights that may be gleaned from the three sub-questions — what, who, and why — are of particular interest and are the focus of this chapter.

The "what" of IT innovation

The first and most critical question is that of the value or use of IT in second or foreign language education. One of the most obvious advantages of computer technology in general and the World Wide Web in particular is that it gives users easy access to a tremendous amount of information. From a language teaching perspective, the web is a wonderful resource for teachers who wish to obtain authentic materials in the target language — and this is particularly true for instructors working in foreign language contexts. ESL or EFL teachers, for example, could download thematically related articles on issues of current interest (the environment, sport, political stories of the day) from, for instance, *The New York Times*, *The London Times*, and *The South China Morning Post*. These input materials could be exploited in an advanced oral discussion class to establish not only the commonalities in how the particular story is reported in these three newspapers but also to highlight subtle differences in editorial position and interpretation. This analysis could subsequently lead to a writing assignment in which students are asked to critically analyze the information they had read and to develop their own position on the issues discussed in the source materials. This template for exploiting authentic language materials could be further complexified by incorporating streaming audio or video files from radio and TV stations that broadcast on the web (the BBC, CNN, or other English-speaking networks), thus incorporating a variety of listening passages into the basic mix of written texts.

The web is also a useful tool for delivering task-based, information gap, or jigsaw activities to learners organized in small groups. For example, to develop further the previous example of how to exploit thematically related articles taken from different newspapers, learners could be divided into three face-to-face oral discussion groups: a *New York Times* group, a *London Times* group, and a *South China Morning Post* group. Each group would be directed to click on the appropriate hyperlink on a webpage that the teacher had set up, and each group could work on appropriate comprehension activities within their own group. In order to ensure that an identifiable product was generated by this phase of small group work, the learners in each group could perhaps be required to summarize their discussions using a bulletin board such as web board. Once the teacher was satisfied that the members of each group had thoroughly discussed and understood their respective articles, he or she could then reconfigure the groups so that the new groups that were formed, each having at least one representative from the original groups. The task of these newly formed groups would then be to analyze how similar or different were the articles that each member had read. Once they had exchanged the necessary information, they could then synthesize their knowledge for use in a later oral presentation and/or written assignment.

If one goes to the trouble of developing such complex pedagogical activities, it is obviously a shame not to be able to revise them and use them again at a later date. Fortunately, the web is also a powerful resource for publishing and archiving one's own materials. Note that, unlike conventional print materials, electronic files can be updated and/or revised very easily. This means that teachers can easily add or substitute new texts to a unit as a story unfolds over time. Alternatively, they can develop existing tasks or add different kinds of tasks to existing materials, or perhaps even change the pedagogical focus of a unit.

As teachers and institutions develop more web-based pedagogical resources, the need to organize such materials in a coherent fashion becomes ever more pressing. Most teachers probably start out by developing their own systems of organization. However, commercial web-based course management software packages such as WebCT or Blackboard are becoming increasingly attractive solutions. Not only do they provide standardized solutions for organizing text, graphics, audio, and video files into a single environment, but they also incorporate potentially interesting additional functionalities into a single package. For example, all these packages provide opportunities for synchronous and asynchronous communication in the form of bulletin boards and class email capabilities, respectively. Furthermore, they enable teachers to develop online quizzes and tests. These tests — which typically employ multiple-choice questions, true-false, matching questions, and other objective formats — are graded by the software as soon as they are submitted by students. This capability means that the turn-around time for providing not only grades but also targeted feedback on incorrect answers is dramatically cut from possibly a few days to a few seconds. An additional benefit of online testing is that this technology frees teachers from the drudgery of grading and potentially enables them to concentrate on the more creative aspects of teaching.

Furthermore, assuming that the answers to test questions have been correctly coded in the first place, the potential for human error in grading is also minimized.

These, then, are some of the advantages of using IT for language teaching purposes. However, we should not be lulled into thinking that IT is the magic bullet that will solve all of our problems as language teachers. For example, in the delivery of the jigsaw tasks via the web, described previously, it is assumed that the groups meet face-to-face in the same computer-equipped classroom. After all, a skillful teacher can exploit the capabilities of the web to organize and deliver large amounts of textual and other information to different groups. However, at the present time, it would be difficult to conduct these kinds of orally-mediated information gaps and jigsaw tasks in a *distributed learning environment* — that is, in an environment in which *The New York Times* group was physically located in the United States, *The London Times* group in Britain, and *The South China Morning Post* group in Hong Kong.

Video conferencing software has made great strides in recent years. However, *latency effects* — that is, the delays that occur between the end of a current speaker's turn and the beginning of the next speaker's turn — are a noticeable artifact of even the best professional software used by local and international news organizations such as CNN and the BBC. It is precisely these latency effects that prevent learners from being able to initiate comprehension checks and other repairs (Long 1996) at the precise, conversationally relevant moment in the unfolding interaction and thus get the comprehensible input they need to learn new language. In summary, although the widespread deployment of Internet II will doubtless go a long way toward resolving these kinds of issues, it seems that the Internet is particularly well suited to developing learners' listening, reading, and writing skills. However, current web-based technology is still technically inadequate to meaningfully develop learners' *conversational* skills in a second or foreign language.

At a more general level of discussion, we should also always remember that access to computers and the Internet varies tremendously around the world. Research universities in the developed world have access to fast, reliable networks that make downloading large graphics or video files a routine task. In the countries of the underdeveloped world, however, as Murray points out (this volume), access to computer technology is either still not available or highly restricted. Furthermore, even when it is available, the hardware and software used may be so slow that downloading even a moderately large text file becomes an exercise in technological purgatory.

The problems associated with unequal access to IT are compounded by the fact that IT is a very young, and therefore still unreliable, technology. In addition, IT is changing at breakneck speed. This means that keeping up with the technology is extremely expensive — thereby further exacerbating problems of unequal access —and that developing a set of widely accepted best practices for users becomes extremely difficult.

We should also acknowledge that the use of IT poses special problems for language teachers that our colleagues in other disciplines (especially the sciences) do not necessarily face. More specifically, whereas chemistry, for example, has a well-defined body of knowledge that has to be mastered by learners, language learning has no content beyond language itself. Traditionally, the content matter of language has been understood to mean its grammar and vocabulary. However, the whole thrust of the communicative language teaching movement during the last twenty to thirty years has been to emphasize the importance of learners being able to use language for communication and not just to learn its formal grammatical characteristics. In this context, for example, we should realize that a language course delivered entirely through WebCT could run the methodological risk of taking several giant strides backwards, by focusing excessively on what this course management software package is able to test in an objective fashion, namely, grammar and vocabulary. In other words, language teachers must beware of developing their courses by what the technology can currently do. Instead, they should ask what they need to be able to do in order to promote language learning and then choose whatever technological solutions are available that enable them to achieve their pedagogical ends.

If we apply a diffusion of innovations (Rogers 1983; 1995) perspective to IT, an important distinction is made in the literature between so-called *primary* and *secondary* innovations (Markee 1997a). Primary innovations include the introduction of fundamental changes in an institution's core educational practices. Examples of such innovations include the development of new foreign language requirements, curricula, teaching and testing materials, and the introduction of associated changes in teachers' methodological behaviors and beliefs about teaching. Secondary innovations, in contrast, are enabling innovations. That is, the function of these innovations is to facilitate and support the implementation of primary innovations. It is important to realize that the use of IT in second or foreign language teaching is a secondary innovation. This is true despite the fact that the use of IT may subsequently entail important changes in the way materials are delivered to learners and in how teachers' methodological practices and attitudes about teaching evolve. A description of the situation at the University of Illinois at Urbana-Champaign exemplifies this point.

The UIUC recently introduced a new foreign language graduation requirement for undergraduates. More specifically, in order to graduate, students must have taken three years of a foreign language at high school or else three semesters of foreign language instruction at UIUC. Now, most undergraduates take Spanish as their foreign language. As a result, the enrollment in Spanish courses is extremely large (at last count, over 2000 students per year). In order to accommodate these numbers, the grammar and vocabulary portions of the courses that fulfill this new requirement are now delivered to learners via Mallard, a web-based course management tool developed at UIUC that is very similar in its capabilities to WebCT. Most importantly from a pedagogical point of view, this solution enabled instructors to focus on communicative tasks during in-class activities. And from an administrative perspective, instructors were able to handle double the number of students they would have taught previously under the old model of instruction, without, however, having to put in extra preparation time for their classes. Indeed, since instructors no longer had to grade grammar and vocabulary tests, on average, the amount of out-of-class preparation time instructors had to do actually went down, as did the contact hours. More specifically, instructors now meet face to face with their students three hours a week instead of four hours. Students use the fourth hour that has been freed up to interact in writing with their peers via an electronic bulletin board. Note that a comparison of students' final grades showed that learners who had taken the web-based version of the courses did as well as, if not better than, their counterparts who had taken the traditional versions of the same courses. Thus, in this example, the new foreign language requirement constitutes the primary innovation, while the use of IT is the secondary innovation which makes the implementation of this requirement pedagogically, administratively and indeed, financially possible.

This distinction between primary and secondary innovations is important. By definition, providers of secondary innovations are viewed as service departments. In the culture of a research university, such departments have less influence than do academic departments. As a result, service departments generally receive fewer financial and human resources than academic departments. As will be shown, this can be extremely problematic for an IT provider such as the LLL, since installing and maintaining computer technology over the long term is an extremely expensive proposition.

The "who" of IT innovation

Many people are potentially involved in implementing innovations. Typically, participants adopt different roles as change agents, adopters (or resisters), implementers, clients, or suppliers (Lambright and Flynn 1980). Note that these roles are not mutually exclusive. For example, implementers and clients are not just passive consumers of innovations; they, too, make decisions about whether they like or dislike innovations, and in this sense they also act as adopters or resisters. Nonetheless, these categories are a useful heuristic device.

More specifically, change agents are the people who initiate change. Thus, in the Spanish as a Foreign Language (SFL) example at the UIUC, a professor in the Department of Spanish, Italian, and Portuguese and the associate provost responsible for undergraduate education functioned as the principal change agents. Change agents seek to persuade potential adopters (or their opposites, resisters) to accept innovations so that these changes become incorporated into an institution's daily practice. In the case of SFL at UIUC, the principal adopter that the change agents had to persuade was the dean of the College of Liberal Arts and Sciences.

Implementers are the people who actually put an innovation into practice. In educational situations, implementers are usually teachers (in this case, Spanish Teaching Assistants), although students (here, UIUC undergraduates seeking to fulfill their foreign language requirements) typically take on the role of clients. Finally, suppliers provide potential adopters with goods and services or provide information about the innovation(s) that are being implemented. In this particular example, we again need to draw on the distinction between primary and secondary innovations. For instance, curriculum designers and materials writers supplied student-clients with the primary innovations of web-based courses, although IT specialists from the LLL provided the enabling technical support for these web-based courses.

For present purposes, I wish to focus on the kinds of issues that confront IT specialists. Clearly, IT specialists need to understand how the technology works and how to fix it when — as it always seems to do — it breaks down. But the crucial variable in the computer support equation is not so much how to support the hardware and software that is used during course delivery. Rather, it is how to help potential users actually *use* this technology as effectively as possible.

This goal raises some interesting issues about the background of instructional designers — those elusive persons — who are familiar with the technology, its pedagogical potential, and the TESOL field. Ideally, these three strands should be combined in one person. In practice, however, these different areas of technological, intellectual, and (perhaps, just as importantly) ultimately *political* expertise and clout that is associated with faculty standing may be divided among different individuals who work together in providing IT support. This was certainly the situation at UIUC. In all cases, however, as I argue later in this chapter, achieving the goal of providing high-quality technical and pedagogical support requires IT specialists to reach out to potential clients by providing them with varied opportunities to develop their technical skills.

The "why" of IT innovation

Some issues relevant to why innovations are adopted or rejected . include the psychological profiles of adopters themselves and the attributes of innovations that either promote or inhibit the adoption of innovations. For present purposes, I focus on the latter category.

Rogers (1983; 1995) has identified at least five attributes that affect decisions to adopt or reject an innovation: (a) the relative advantages to be gained from adopting an innovation, (b) compatibility, (c) complexity, (d) trialability, and (e) observability of the innovation. Again, these terms are explained in reference to SFL at UIUC.

The notion of an innovation's relative advantage is crucial in the adoption process. As already noted, some of the advantages for instructors who taught the web-based sections of these courses were substantial. They included a reduction in both out-of-class preparation time and contact hours, relief from the more time-consuming and mechanistic aspects of language instruction (test preparation and grading), and the ability to focus on communication-oriented tasks during face-to-face instruction. However, in exchange, teachers now became responsible for forty students per semester instead of twenty.

From the perspective of the university administration, the solution of using the web to deliver SFL instruction to large numbers of undergraduates was also attractive, though for different reasons. Although the up-front financial costs were high — some US$600,000 — these were in fact far lower than the costs that would have been incurred if extra language instructors had been hired to meet the increased demand for Spanish instruction. In addition, the benefits associated with improving the infrastructure needed to support the delivery of web-based SFL courses — a high-speed network for the Foreign Languages Building, completely new computer laboratories for the LLL, and greatly improved technical support services — could automatically be made available to other language departments housed in the Foreign Languages Building. Indeed, these improvements held out the possibility of promoting the use of web-based instruction in other departments, not just in foreign language teaching but in literature and culture studies also. Finally, a further benefit that has emerged over time is that this model of web-based instruction has been adopted by a number of other universities in the United States that were also looking for a viable solution to the problem of how to cope with large enrollments in Spanish.

The influence of the other factors identified by Rogers (1995) in promoting the adoption of web-based instruction is also discernible, although issues of the compatibility, complexity, trialability, and observability of this innovation are undoubtedly less important than its relative advantages and disadvantages.

More specifically, the issue of *compatibility* has to do with how similar or how different an innovation is from present practice. As Stoller (1992, 1994) has shown, innovations that are moderately innovative are the most likely to be adopted. In contrast, innovations that are too similar or too different from potential adopters' current practices are unlikely to be adopted. It seems that the web-based Spanish courses were perceived by users to be sufficiently different from traditional instruction, particularly in their exploitation of online testing, to count as new, while not being so different to be disorienting.

In this respect, the relative ease of use — or lack of *complexity* — of modern browsers and web-based tools is an important factor to be considered. Ten years ago, accessing information on the web was a daunting process. However, because of the advent of browsers such as Internet Explorer, surfing the web has become a form of entertainment that many people — particularly teenagers — engage in as a matter of routine. Thus, the overwhelming majority of students enrolled in the web-based Spanish courses could be counted on to bring substantial skills and prior experience in using the web before they even enrolled in the courses. Furthermore, since Mallard, the course management software package used to deliver the SFL courses, is a web-based tool, it is *platform-independent*. This means that Windows and Macintosh users can have essentially the same computer-mediated language learning experience, irrespective of their hardware preferences.

The issue of *trialability* has to do with whether potential adopters can try out or assess an innovation before they commit themselves fully to adopting it. In the case presently under discussion, for example, the dean's decision to adopt web-based instruction for Spanish was also influenced by the fact that the professor-change agent who promoted this idea had already gained substantial experience in developing web-based courses in Italian, but for comparatively small numbers of students. The dean could therefore be confident that many of the problems that affect computer-based instruction had already been successfully resolved before he committed himself to a "solution" that might potentially cause more problems than it solved.

Finally, the issue of *observability* is also important in persuading potential users that it might be worthwhile to adopt an innovation. In the case of the web-based Spanish courses, for example, the fact that the LLL's main computer site was redecorated and completely re-equipped with state-of-the-art equipment to service this project was a change that was very visible to potential users. Furthermore, as enrollment in the web-based Spanish courses grew, the numbers of people using this site grew from a few individuals a day to several hundred a day. This not only made the SFL courses highly visible to undergraduate students who had not yet enrolled in a language course to fulfill their foreign language undergraduate requirement, but it also made the concept of computer-mediated language, literature, and culture instruction highly observable to any faculty member, teaching assistant or student in other departments who happened to visit the LLL's computer laboratory.

POSITIONING AN IT CENTER TO SUPPORT CURRICULAR INNOVATION: LESSONS FROM UIUC

As purveyors of secondary innovations, the independent existence of IT centers cannot be justified. The success of IT centers such as the LLL is dependent on the mission of such units being closely linked to curricular initiatives that are important priorities for the wider institution. If this link does not exist, it is highly likely that IT centers that attempt to "go it alone" will be marginalized.

In the UIUC context, the curricular initiative to which the renovation of the LLL was explicitly tied was the implementation of a new undergraduate foreign language requirement and the consequent development of new SFL courses for learners who wished to fulfill this requirement by taking Spanish. But clearly, an opportunity existed to promote the use of computer technology in all of the departments in the Foreign Languages Building, not just in Spanish. This resulted in the development of two important initiatives by the LLL that were closely interrelated: a consultancy service to help faculty write grant proposals to obtain equipment and training from the university, and the Outreach Program. These two secondary innovations are described next.

The consultancy service

For many years, UIUC has funded small-scale faculty projects involving the use of computer technology. Historically, the bulk of funding awarded by the committee that makes these awards has tended to go to faculty in the hard sciences and agriculture, largely because faculty in these areas tend to be more computer literate than their counterparts in the humanities. Thus, few faculty from the departments in the Foreign Languages Building bothered to apply, and even fewer would receive funding — one award per year being usual. Consequently, the computer equipment that faculty in the Foreign Languages Building had access to in their offices tended to be old and out-of-date, which further discouraged experimentation with web-based technologies.

As the incoming director of the LLL, I was appointed to serve on the committee that made the awards for these small-scale faculty projects. This meant that I was well positioned to develop a good understanding of the committee's priorities. Thus, setting up a consultancy service that provided advice to potential grant writers in the language departments on how to write their proposals in such a way that their chances of being funded would be maximized was a natural step. This service has proved to be very successful. For example, in the first year that this service was set up, seventeen faculty from the language departments applied for equipment and training and fourteen were funded.

One of the most successful projects that was funded in this way involved the Department of East Asian Languages and Cultures that developed a web-based course in Mandarin Chinese for heritage language learners. This course was technically modeled on the example of the Spanish courses. More specifically, it successfully used the online quizzing capabilities of Mallard to help Chinese-American students who spoke but did not read or write Chinese develop their literacy skills in this language. The LLL's role in this process included not only helping the professor write his grant proposal but also demonstrating what kinds of exercises had been used by the Spanish programmers, thus helping the Chinese team decide how to develop their own solutions for the heritage Chinese course. The demonstration aspect of the consultation was carried out by the Outreach team.

The outreach program

For a variety of reasons, which we need not go into here, prior to my appointment as director of the LLL, this unit was grossly underfunded and understaffed. More specifically, there was only one person available to provide technical support for a building that has approximately 100 faculty and 250 teaching assistant users. Even then, providing technical support for users was not part of the official job description of this individual, who was really only responsible for supporting the building's network. Thus, any technical support that was provided was strictly ad hoc. Fortunately, the dean of the College of Liberal Arts and Sciences recognized that technical support had to be part of the maintenance package that was needed to ensure the success of the new Spanish courses. Consequently, two technical support staff were permanently allocated to the LLL and a third person was lent for a period of two years.

This increase in the numbers of technical staff obviously represented a very important improvement in the ability of the LLL to provide meaningful technical support to users. However, it was still not enough according to the sheer volume of support work that had to be accomplished. This was particularly true in the area of encouraging more faculty and other instructors to engage with computer technology as a resource for teaching. Unfortunately, the traditional technical support duties (for example, fixing non-functioning email, printers, and other common problems of this kind) quickly threatened to overwhelm the LLL's capacity to promote the wider use of computer technology, which was by far the more important task. A creative solution was clearly needed which enabled the LLL to fulfill both functions. It is out of this necessity that the Outreach program was born.

The solution that was eventually devised involves a close partnership between the associate provost responsible for undergraduate education, who provided the funding for the Outreach program, and the LLL, and client departments. More specifically, the Outreach program is staffed by graduate students from ten participating departments. These individuals work for twenty hours a week as fifty percent research assistants. Each student's home department funds half of the appointment, and the LLL funds the other half. The LLL provides office space, computer equipment, and training for the research assistants. In addition, one of the LLL's full-time consultants acts as the program's overall co-ordinator. In this way, departments have an important say in who works for them, and this enhances the departments' sense of ownership of the Outreach program.

The Outreach research assistants have two functions. For four hours a week, they provide basic technical support for Foreign Languages Building faculty and staff. This arrangement enables the LLL's full-time technical consultants to function as a second line of defense for those more complex problems that Outreach research assistants cannot easily solve. At the same time, this arrangement gives the full-time consultants the time they need to maintain the LLL's laboratories and to attend to their other duties. For the remaining sixteen hours a week that is associated with their Outreach appointments, research assistants work on a variety of computer-mediated curriculum development projects for their home departments.

Thus, the basic idea here is that a faculty member who wishes to put up a course on the web can concentrate on developing course content, while the department's Outreach research assistant provides the necessary technical assistance for the task at hand. Of course, this division between content and technological support is never that clear-cut. In some instances, for example, a research assistant may know a lot about a given content area and may therefore be entrusted by the faculty member to help develop content for a course. Nonetheless, the basic division of labor described here is valid.

These kinds of partnerships have resulted in five applied linguistics courses and over forty units of task-based ESL materials being put on the web in the Division of English as an International Language, and in web-enhanced language materials of various kinds being developed in the French and German departments. Other projects in the pipeline include the development of an online data base of several hundred slides that are to be used in a general education course on Classical civilization, to be followed by similar projects in French, Italian, and Spanish culture courses.

Last but not least, the Outreach program is actively engaged in disseminating information about IT by staging public demonstrations of projects in which it has participated and by organizing a growing program of training workshops. Users may now sign up for training workshops on a variety of software packages (including, for example, *Dreamweaver*, a package which is used to develop web pages) or *PowerPoint* (which is used to present information during lectures). As a result of this training, users are being given the skills that, in the long term, enable them to develop their course materials on their own without having to depend on the continuing help of Outreach program. To date, some 100 to 150 users have taken advantage of these training opportunities during the last year.

These Outreach activities have significantly enhanced the quality of technical support that is now available to users in the Foreign Languages Building. Furthermore, they have helped to institutionalize the use of computers in the everyday teaching practices of many language, literature, and culture teachers in this building.

Because of the decline in the US economy from 2002 onwards, the University of Illinois suffered dramatic declines in its budget. As a result, the LLL was absorbed into a larger IT support structure within the College of Liberal Arts and Sciences. This new IT unit now provides computer support for the entire college, not just the foreign language and literature faculty. This enables important economies of scale in a centralized unit being in charge of purchasing new equipment for all computer laboratories, not just the LLL. However, the achievements of the LLL's Outreach program during the five years of its existence have been recognized, to the extent that its aims, organizational structures and procedures have been adopted by, and integrated into, the larger program's services. Thus, even though the LLL no longer exists as a separate IT unit in its own right, from a diffusion of innovations perspective, its influence continues under the guise of a larger, more centralized provider of IT.

As I have noted on several occasions in this chapter, however, developing an adequate IT infrastructure is an enormously costly enterprise. It is easier to justify replacing machines in public laboratories every three years or so than it is to secure recurring funding for the technical personnel who are needed to maintain initiatives like the Outreach program.

This is not altogether surprising. People who have the kind of background in applied linguistics, second, or foreign language teacher training and IT that is necessary to provide the hands-on supervision for an initiative such as the Outreach program are rare. They are also very expensive. In the UIUC context, for example, an individual with this kind of background (who may or may not have a PhD) can expect to earn as much as a new assistant professor in the humanities. And in the context of the commercial sector, the same technical consultant could easily earn twice his or her university salary by working for a private company.

Thus, it is inherently a very difficult proposition for universities to attract and keep people with these kinds of skills. Consequently, irrespective of the specific details of the situation that exists in a particular institution, I would identify the problem of providing *ongoing* support for end users as the most important problem for universities to solve. Finding a viable solution to this problem is obviously crucial if individuals who have no inherent interest in IT are to be persuaded that it is worthwhile using such technologies over the long term. However, it is not all obvious how to develop these solutions or what these solutions might look like.

CONCLUSION

In this chapter, I have attempted to lay out how we might conceptualize the use of IT to promote curricular innovation in second or foreign language education. In order to do this, I have framed my discussion of what I consider to be the most important issues that confront IT centers such as the University of Illinois Language Learning Laboratory, within the theoretical context of the diffusion of innovations literature. More specifically, I believe that, whether one is a teacher, an administrator, or a provider of IT services, it is especially important for us to understand how the variables that I have labeled as *who-*, *what-*, and *why*-related factors affect the ultimate success or failure of IT centers. We need to understand that many different kinds of people playing a variety of social roles are involved in making innovations work. We also need to understand the kinds of things that IT can do well in the context of second and foreign language teaching, and what it cannot do well, at least at the present time. Furthermore, we need to understand the essentially ancillary role of IT in second or foreign language education. Finally, I have argued that we need to understand why people adopt or reject innovations according to of five attributes that either promote or inhibit the spread of an innovation. These attributes include the relative advantages of an innovation, its complexity, its compatibility, its trialability, and its observability.

From a more practical perspective, I have described the LLL's attempts to engage with potential users through its consultancy service and Outreach program. From this discussion, two main lessons emerge:

1. From an organizational standpoint, the success of IT centers such as the LLL is dependent on the mission of such units being closely linked to curricular initiatives that are important priorities for the wider institution.
2. Providing *ongoing* support for end users is crucial if individuals who have no inherent interest in IT are to be persuaded that it is worthwhile using such technologies over the long term. What such solutions look like, however, is not obvious.

In addition, I would like to suggest that the kinds of issues that face IT centers such as the LLL are not language-specific. For this reason, I have tried to provide examples of projects of computer-mediated curricular innovation that targeted a number of quite different languages. Indeed, I maintain that it is also advisable for IT centers to position themselves as providers of IT services, not just to its traditional clientele of people interested in second or foreign language instruction but also to literature and culture specialists. Only then will it be possible to obtain the kinds of economies of scale that are necessary to justify the provision of the complex, and extremely costly, infrastructure that is needed to help language specialists use IT on an everyday basis.

NOTE

1. For a comprehensive list of references on the management of curricular innovation, see the references in Markee (1997a). More recent references or references not listed in Markee (1997a) include: Kennedy (1999); Kennedy, Doyle, and Goh (1999); Li (1998); Markee (1997b); Rea-Dickins and Germaine (1998); Salaberry (1999); Stoller (1997); Wall (1996).

REFERENCES

Cooper, R. (1989). *Language planning and social change*. Cambridge: Cambridge University Press.

Kennedy, C. (ed.) (1999). *Innovation and best practice*. London: Longman.

Kennedy, C., Doyle, P., and Goh, C. (eds.) (1999). *Exploring change*. London: Macmillan Heinemann.

Lambright, W. and Flynn, P. (1980). The role of local bureaucracy-centered coalitions in technology transfer to the city. In J. Agnew (ed.) *Innovation research and public policy, Syracuse Geographical Series*, No. 5 (pp. 243–82). Syracuse, NY: Syracuse University Press.

Li, D. (1998). It's always more difficult than you plan and imagine: Teachers' perceived difficulties in introducing the communicative approach in South Korea. *TESOL Quarterly*, *32*, 677–703.

Long, M. (1996). The role of the linguistic environment in second language acquisition. In W. Ritchie and T. Bhatia (eds.) *Handbook of second language acquisition* (pp. 414–68). New York: Academic Press.

Markee, N. (1993). The diffusion of innovation in language teaching. *Annual Review of Applied Linguistics: Issues in Second Language Teaching and Learning, 13*, 229–43.

Markee, N. (1994). Curricular innovation: Issues and problems. *Applied Language Learning, 5*, 1–30.

Markee, N. (1996). Making second language classroom research work. In S. Gass and J. Schachter (eds.) *Second language classroom research: Issues and opportunities* (pp. 180–218). Hillsdale, NJ: Lawrence Erlbaum.

Markee, N. (1997a). *Managing curricular innovation*. New York: Cambridge University Press.

Markee, N. (1997b). SLA research: A resource for changing teachers' professional cultures? *Modern Language-Journal, 81*, 80–93.

Rea-Dickins, P. and Germaine, K. (1998). *Managing evaluation and innovation in language teaching: Building bridges*. Harlow: Longman.

Rogers, E. (1983). *Diffusion of innovations* (3rd ed.). New York: Free Press.

Rogers, E. (1995). *Diffusion of innovations* (4th ed.). New York: Free Press.

Salaberry, R. (1999). The research-pedagogy interface in L2 acquisition: Implications for language program directors. In L. Heilenman (ed.) *Research Issues and language program direction* (pp. 39–59). Boston: Heinle and Heinle.

Stoller, F. (1992). *Analysis of innovations in selected higher education intensive English programs: A focus on administrators' perceptions*. Unpublished doctoral dissertation, Northern Arizona University.

Stoller, F. (1994). The diffusion of innovations in intensive ESL programs. *Applied Linguistics, 15*, 300–27.

Stoller, F. (1997). The catalyst for change and innovation. In M. Christison and F. Stoller (eds.) *A handbook for language program administrators* (pp. 33–48). Burlingame, CA: ALTA Book Center.

Wall, D. (1996). Introducing new tests into traditional systems: Insights from general education and from innovation theory. *Language Testing, 13*, 334–54.

4

Building a Bridge: A Case Study in Crossing the Innovation Chasm

Chris Corbel

INTRODUCTION

Although the English language teaching literature contains many examples of successful innovations, the reasons why some innovations are so successful and so many others fail are rarely addressed (Fullan 2003). Carlopio (2003) suggests that there are macro- and micro-level reasons for innovation success. At the macro level are issues such as organizational culture and structure. At the micro level are issues such as the perceived characteristics of the innovation, and the characteristics of individuals. Goldsmith, d'Hauteville, and Flynn (1998) and Johnson, Donohue, Atkin, and Johnson (2001) have concluded that innovativeness can be identified as a valid construct at the individual level.

A well-established view of the adoption cycle by individuals for any technological innovation is the diffusion model developed by Rogers (1983). Rogers suggests that the distribution of new adopters of a technology over time will be approximately normal, any individual falling into one of five distinct categories (innovators, early adopters, early majority, late majority, and laggards) according to his or her social and psychological characteristics (p. 243). Moore (1991) examines these categories from the perspective of an external vendor or internal change agent wishing to introduce a technological innovation. He identifies three markets for any technological innovation: early, mainstream, and late.

Geoghegan (1994, 6–7), expanding on Rogers (1983) and Moore (1991), describes the five categories of adopters as follows. Innovators (two to three percent of adopters) are interested in the technology as much as in the application and may be members of broad, cross-disciplinary networks of like-minded individuals. Early adopters (about ten percent of adopters) are interested in the possible application of the technology to professional tasks. They are project-oriented risk takers, willing to experiment, and are reasonably self-sufficient. The early majority is more pragmatic, wanting proven applications. Their networks are more vertically discipline based, rather than cross-disciplinary. The late majority are similar to the early majority, though perhaps less comfortable with technology. The last fifteen percent, the laggards, may never take up the innovation at all.

The diffusion model implies a smooth transition between stages of adoption, but Moore (1991) suggests that this may not be the case in practice. He identifies a potential "chasm" between uptake by the early market — *innovators* and *early adopters* — and the mainstream market — *early and late majorities* (p. 17). "Crossing Moore's Chasm", as it has come to be called, is thus a potential issue for anyone with responsibility for the introduction of a technology innovation.

Geoghegan (1994) suggests four reasons why the chasm, or "gap" as he calls it, has not been bridged in higher education in the United States: (a) ignorance of the gap, as a result of an assumption about homogeneity of interest and attitudes among staff; (b) an informal "technologists' alliance" between innovators, early adopters, information technology support units, and external vendors that leads to an underestimation of the support needs of mainstream staff; (c) a resulting alienation of the mainstream from the highly visible innovative minority; and (d) the lack of a compelling reason for mainstream staff to get involved with any particular product in a seemingly never-ending cycle of technological change (pp. 9–11).

Geoghegan (1994) suggests four steps that need to be taken to ensure the gap is bridged as follows: (a) recognition of the mainstream group as distinct and different and not easily brought around to the visions of the innovators and early adopters; (b) recognition that mainstream support needs to be provided through discipline-based peers rather than horizontal networks and technology alliances; (c) provision of a compelling reason to adopt the innovation; and (d) provision of an institutional commitment (pp. 11–4).

Geoghegan (1994) describes the introduction of the Internet into learning in the early 1990s. Since that time there has been an enormous increase in online learning initiatives across all education sectors. Yet, in my view, the issues involved in crossing the chasm, or bridging the gap, are as important now as they were in the early 1990s. This view is based on experience managing the development and maintenance of three online learning environments: the Technical and Further Education (TAFE) Virtual Campus (VC), which provides a platform for hundreds of learning modules offered by dozens of providers to thousands of students of vocational education and training in the state of Victoria, Australia; aclEnglish, an extensive range of online and on-paper courses, resources and services; and the Virtual Independent Learning Center (VILC), which offers a range of English Language Teaching resources to teachers and learners at over 150 institutional sites throughout Australia and Asia.[1]

In spite of the potential need to bridge the innovation chasm in order to ensure the effective uptake of new online learning opportunities by large numbers of teachers, there is relatively little treatment of the issue in the literature on IT and educational change. The Interactive Learning Center (1998) at the University of Southampton is a rare exception, offering techniques for specifically addressing related issues. This chapter therefore seeks to add to our understanding of technology innovation uptake in education by presenting a case study of an online learning project that was explicitly intended from the outset to involve as many mainstream users as possible. Although it was not specified in the project brief, implicit in reaching the project goals was the need to bridge Moore's chasm.

In the following section, the development and implementation of a simple online learning environment, the VILC, is described. In the final section, the development is analyzed from the perspective of the four steps recommended by Geohegan (1994) as necessary to bridge Moore's innovation chasm between the innovators and early adopters and the early and late majorities of users.

BUILDING THE BRIDGE ACROSS THE INNOVATION CHASM

In December 1996, Adult Migrant Education Services (AMES) Victoria, a large specialist adult ELT organization with over 400 teaching staff, successfully tendered for funding to develop materials for English teaching online. The funding was made available through the state Office of Training and Further Education (OTFE) to develop online learning in TAFE institutions throughout the state of Victoria, Australia.

The project developer had already successfully managed an earlier organization-wide curriculum innovation (Bottomley, Dalton and Corbel 1994) and was aware of the issues involved in implementing broad organizational change. He had also carried out earlier surveys of staff both inside and outside AMES Victoria in relation to computer use (Corbel 1996; 1998). The skill levels and needs of staff had thus already been broadly identified. A clear area of need that had emerged was the effective use of the Internet in teaching.

The project was managed by the AMES Computer Literacy Center, a small (two-person) unit that was established in 1996 to build on emerging opportunities in online learning. Although the focus of the Computer Literacy Center was on training, both staff development and online resource creation were seen as important growth areas. Accordingly, it was proposed that teachers, not computer specialists, would develop content for the project.

The online materials development project began at the start of 1997 when staff throughout AMES were invited to participate. Interested staff were not to be recruited out of existing groups of CALL specialist teachers but rather from among those who were experienced, interested, and were comfortable working in what would be a relatively unstructured project. Seven teachers were involved in the initial year, working one day a week on the project.

The main goal of the project was to produce a collection of teaching materials to be delivered via the Internet. The focus was on developing online resources from an existing set of distance learning course materials used in the national Adult Migrant English Program (AMEP). Within the first few weeks however, it became apparent that the time taken to develop the materials would be greater than anticipated because of the complexity of the target materials. The original distance learning resources were seen to be too difficult for non-specialists to revise, particularly because the materials made extensive use of embedded audio and video files within a complex structure. It was also realized that such complex resources would put demands on potential users' ability to access the Internet, many of whom had limited access at best. Simpler text-only materials, it was decided, would be easier for teachers to produce and easier for students to access.

Since another goal of the project was to develop realistic and sustainable materials development processes for teachers throughout AMES, the project managers reconsidered their initial use of distance learning materials. It was decided to focus efforts on the essentials of what teachers do already and how that could work online. Based on the familiar communicative language teaching paradigm, it was argued that teachers would most likely want to work with authentic texts and other realia, to create activities in order to meet specified learning objectives. Since the whole of the Internet is made up of authentic texts and messages, the project was then re-oriented towards creating a means for teachers to frame that realia under a curriculum goal in a way that emulated what they already did in the classroom. Thus, the teachers would not create content; they would create learning tasks based on existing content.

Once the concept of utilizing authentic online content was adopted, project participants spent several weeks discussing and refining a task framework that would be applicable for the target curriculum, the Certificates of Spoken and Written English (CSWE). Individual project members then identified a topic or theme and began searching the Internet for suitable content. The project participants spent the remainder of the year writing tasks to meet the outcomes of the CSWE.

At the beginning of 1998, the project management sought internal funding to further develop the resource that is now called the Virtual Independent Learning Center. The name makes reference to the many Independent Learning Centers (self-access centers) that exist throughout AMES and in various forms at a number of other educational institutions. One typical feature of independent learning centers throughout AMES is racks of plastic folders, each one containing a piece of realia and activities created by teachers. These are organized according to curriculum levels. Each VILC task, therefore, was seen to be the virtual equivalent of one of these tasks.

A business plan was written to identify ongoing revenue streams for VILC, in order to further develop the site without reliance on internal funding. At the end of 1998, a subscription-based business model was put into place. At the beginning of 1999, the Commonwealth Department of Immigration and Multicultural Affairs (DIMA) agreed to meet the subscription fee for all AMEP providers in Australia. For its part, AMES agreed to continue to subsidize one staff member part-time.

The AMEP sponsorship led to the immediate establishment of a large community of users, which was carefully supported and nurtured. The number of task writers rose, as teachers discovered how simple it was to get started. There was also the beginning of an expansion in the number of curricula, with the setting up of an industry-collection of tasks managed by the AMES Victoria Workplace Communications Unit. By the end of

1999, VILC was well established throughout Australia. In 2000, a new content stream was introduced, based on Australia's Special Broadcasting Service (SBS) radio news — which is intended for Australians from diverse language backgrounds — as well as content that had been funded from several other sources.

VILC content is now contained within three broad categories — Realweb, Easynews, and Checkout. *Realweb*, the original component of the VILC project, contains hundreds of language learning tasks on dozens of topics for leading ELT curricula. *Easynews* provides learning activities based on specially created weekly radio news items. *Checkout* provides a collection of learning and information resources.

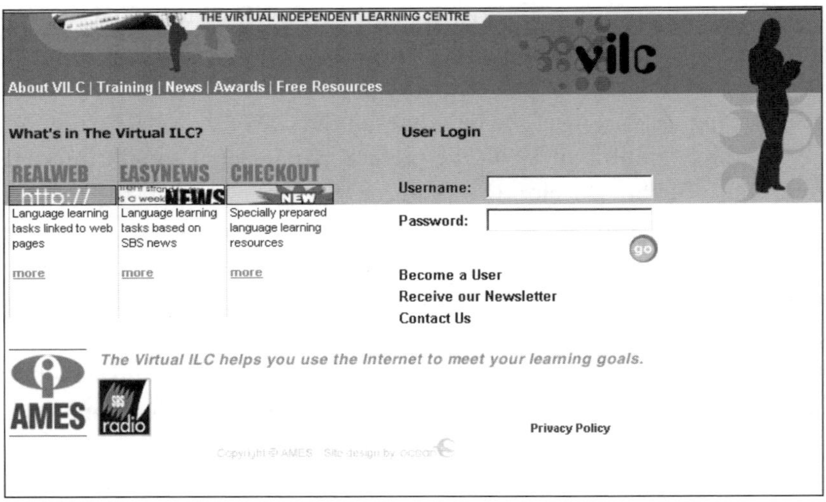

Figure 4.1 VILC homepage

The largest strand in Realweb is the CSWE, a collection of over 500 language learning tasks created by more than sixty language teachers from around Australia, based on popular Australian ELT curricula.

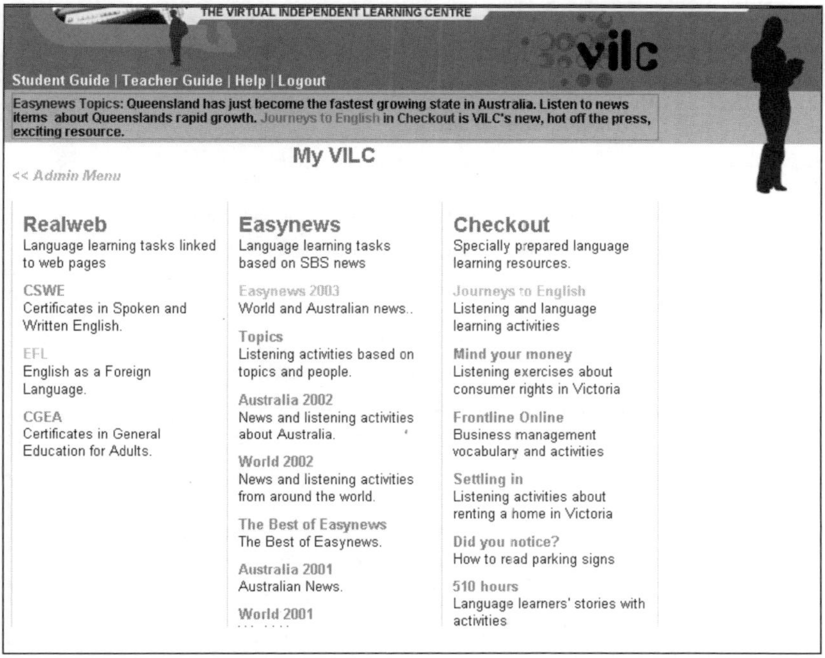

Figure 4.2 VILC contents page

In *Realweb*, teachers create learning tasks aimed at specific curriculum outcomes, based on the "realia" found at other websites, the process described previously. A paper-based equivalent of this would be a teacher creating a worksheet to accompany a chapter from a textbook, a brochure, or some other piece of realia. The teacher does not have to write the book or create the realia but simply "frames" the realia for a particular curriculum goal. In *Realweb*, all tasks are based on websites that teachers have discovered but not created themselves. For example, one teacher has created a series of tasks based on the Immigration Museum of Melbourne's site.

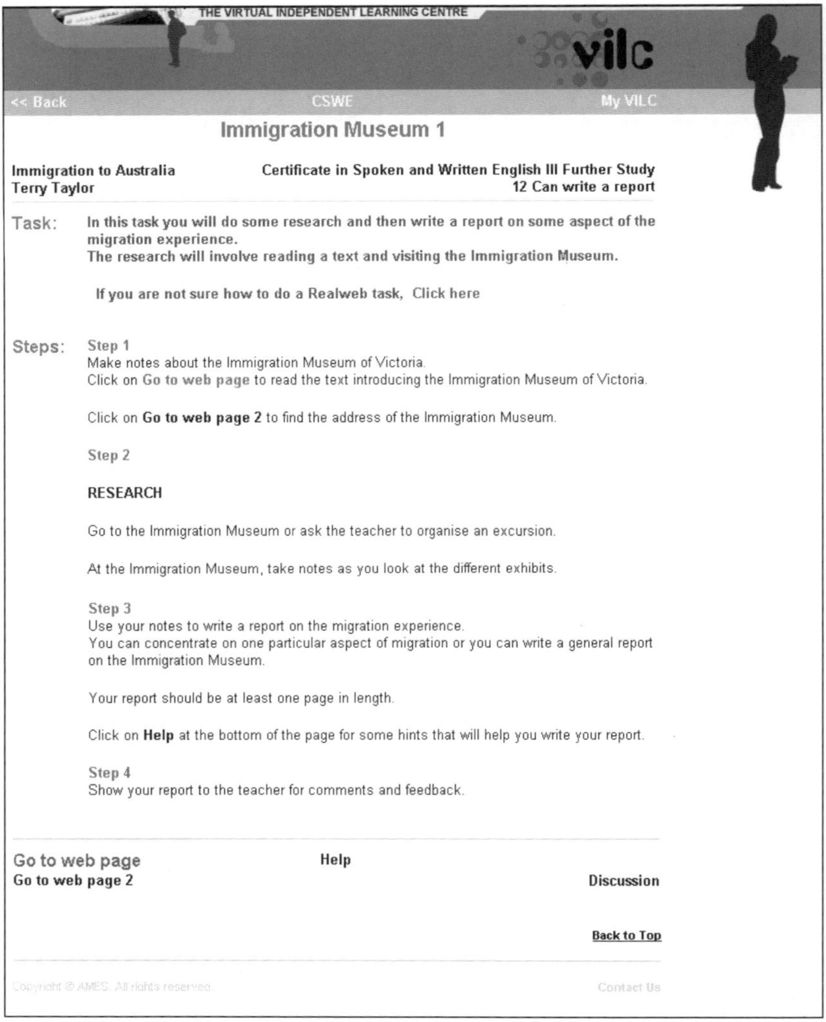

Figure 4.3 A typical Realweb task

The student reads the task and then clicks on *Go to Webpage* to access the Museum's site. Using the *Back* and *Forward* buttons on the browser allows the student to switch between the site task and the target site. The teacher created the task by simply filling in an online form with a word processor. Teachers are thus able to create tasks without having to learn Hypertext Markup Language (HTML), though if they wish, they can use HTML in their task creation.

From the outset, the goal of *Realweb* was to work in a way that would be within the capabilities of any teacher, within the constraints of his or her existing work conditions. By distinguishing between task creation and content creation, *Realweb* allows teachers to add value online, just as they do in person and on paper, by "re-intermediating" (Tapcott 1999) between a resource and a learner to shape the way the learner engages with the resource for an educational goal. The same process is at work when teachers create learning activities based on paper-based realia.

This approach may be considered inconsistent with the freedom of the individual to surf the Internet, discover resources, and engage with them in relatively unstructured ways. Such an arrangement may be appropriate when access and resources are unlimited. It may be appropriate in circumstances that lend themselves towards a project-based approach that emphasizes the social dimensions of language learning (Debski and Gruba 1999; Parks, Huot, Hamers, H.-Lemonnier 2003). However, when teachers and learners have limited access to the Internet, and curriculum goals have to be met within a certain time frame, *Realweb* provides an additional option for teachers and learners. Allowing many teachers each to focus on their own area of interest ensures that a wide range of topics is provided to learners, far beyond what any individual teacher could provide and perhaps beyond what learners could find on their own. This is not to say that learners should not learn to search the Internet, only that they should not have to search it in order to use its resources for learning.

A further goal of the VILC project has been sustainability. Only a small proportion of education projects that receive funding result in products that have a life after the project funding has finished. Websites in particular are often treated as products that are finished when the project is over, whereas they are in fact more like services that need to be maintained and continuously developed. The successful sites often depend on the efforts of an enthusiastic individual, possibly supported by an organization. But the individual may burn out, or the funding may no longer be available. When committing time to learning a new website, and integrating it into a teaching program, a teacher wants to be sure it will still be there next week and next year.

VILC is based on a business plan as well as an educational plan. Access to the tasks developed with the original funding was offered free of charge for the first year after the project finished. At the same time, new tasks were developed. These were made available to educational providers for a low subscription fee that offered unlimited access for a year for their teachers and learners. This was a very simple mechanism that made it easy for potential users to decide whether or not to take up the offer.

The subscription fees help defray the costs of the two staff members who manage the maintenance and development of new tasks for all the strands. Although teachers are not paid directly for tasks in Realweb, their employers are encouraged to support the process and to see task development as part of professional development in online learning. Initial training is offered at cost to assist in this process. All those involved — task writers, the educational institutions that support them, the organizations that sponsor access, and the organizations that provide development funds — are clearly acknowledged throughout VILC.

CROSSING THE CHASM?

To what extent has VILC built a bridge over the innovation chasm and provided a realistic opportunity for a wide range of teachers, early mainstream as well as innovators and early adopters, to work online? Usage has risen steadily over the three years of its operation. For example, there were nearly 5,000 logins in October 2000 compared to 1,500 a year earlier. Over 150 schools, colleges, and universities subscribe from around Australia. Although it is not possible to track usage by individual teachers, VILC is clearly an increasing part of many teachers' work. However, because the number of VILC subscriber organizations is continuing to expand, it is possible that it is the innovators and early adopters at the new user organizations rather than the early mainstream staff at existing user organizations, who account for the increase in usage. Further research is necessary before a clearer picture emerges.

To what extent has *Realweb* task writing, as opposed to VILC usage, become part of teachers' everyday work? At the end of 2000, a detailed evaluation was undertaken (Corbel, Gruba, and Enright 2002) that highlights a number of key points.

Although there were over sixty task writers, not all were continuing to create tasks, and some have dropped out of active task creation. There were a number of reasons, primarily relating to time, access, and roles. Most task writers said they simply didn't have time to write tasks. Teachers work within an extremely full and uniquely (for professionals) constrained environment. Typically, the addition of something new is not balanced by the removal of an existing duty or responsibility. Internet access remains an important issue. Although teachers have email accounts, they may share access with several colleagues and may not have computers on their desktops. If they have access at home, it may be slow. In retrospect, expecting teachers to build online work into existing work structures in a short space of time was optimistic. Although VILC usage is likely to become part of many teachers' work, the hope that VILC task creation would slip easily into individual teachers' work was clearly unrealistic.

At a broader level is the issue of whether the development of *Realweb* tasks should be the work of generalist teachers or specialist teachers. Or, to put it another way, should all teachers be expected to be able to create tasks (or indeed to work online)? In the case of the *Easynews* tasks, one teacher is assigned to task creation and paid for the work. *Realweb*, in contrast, is intended to be the work of a community of task writers, who would benefit professionally by being members. The nature of teachers' work, in person, on paper, and online, is emerging as a critical issue in ensuring commitment to engagement with online learning by teachers in the early majority category (see Corbel 1999, in press, for a more detailed discussion of possible teacher roles).

To what extent can the uptake of VILC be linked to the steps recommended by Geoghegan towards bridging the innovation chasm? As outlined previously, four steps were recommended. The first step was the recognition of the mainstream as a distinct group, who are motivated differently from the early adopters and for whom the early adopters may not serve as realistic role models. A feature of the project has been to ensure that the original project participants and the subsequent task writers are not necessarily members of staff who would be seen by colleagues as technology evangelists. That is, they are not members of the technologists' alliance but are ordinary teachers who are prepared to try something new. Although the project teachers had some of the characteristics of the early adopters, the task writers are more of a mix between early adopters and early majority. They present an achievable role model to colleagues as users of the new technology.

The second step in bridging the chasm was to recognize the vertical orientation of the early and late mainstream staff groups: that is, the fact that their primary networks are with colleagues in their subject area rather than with individuals in other areas who share an interest in IT. Throughout the project, project members and task writers have remained staff members at teaching centers. They do not necessarily have specific roles relating to IT at their workplaces, although some of them do have local support responsibilities. Since most VILC user organizations specialize in ELT, staff at teaching centers have a strong subject area orientation. The project arrangements can thus be seen as acknowledging this vertical integration of mainstream staff. The majority of staff are primarily connected with colleagues teaching the same subject area and depend on peer support that is likely to come from a more skilled colleague, who remains a generalist teacher rather than an IT specialist. VILC users are also able to seek informal support from colleagues who have adopted the task-writer role.

The third step in bridging the chasm was to demonstrate a compelling reason to take up the innovation. VILC now offers two main services, both of which offer clear advantages to users over previous resources. *Realweb* allows teachers and learners to use and develop basic Internet skills and use a wide range of resources in a way that reduces preparation time for teachers and ensures a clear curriculum focus for learning tasks. *Easynews* allows the use of current listening material in a format that does not require high bandwidth. By focusing on what teachers already do with audio- and print-based resources and by using an existing resource, the Independent Learning Center, as a guide for ways of working, teachers have been able to see the advantage of the innovation. Neither resource has severe side-effects which might make its use less attractive, although some users have reported their IT departments have questioned the increased costs of Internet access accompanying the increased use of sound files in *Easynews*. This shows the need for any technological innovation to involve key members of the "technological alliance" (IT departments, innovators and early adopters and vendors), or run the risk of losing key organizational stakeholder support.

The final step in bridging the chasm was institutional commitment. VILC benefited from institutional commitment from three sources: the funding agency that supported its development; the host organization, AMES, which continued to support it with internal funds after the development funds were finished; and the Department of Immigration and Multicultural Affairs, which allowed for a critical mass of users to be established, through sponsoring access for educational institutions delivering the AMEP around Australia.

Arguably, the VILC project has addressed the four steps identified by Geoghegan as central to crossing the innovation chasm, albeit indirectly. Although the project team did not set out to do so, the steps were addressed as a result of other aspects of the project's approach. In particular, the project's goal of maximizing usage by involving as many teachers as possible in the development of content, and its strategy of focusing on staff who were not seen as technological evangelists, can both, in retrospect, be seen as meeting Geohegan's suggestions.

Clearly, further work needs to be done with individual teachers to ascertain which of the adopter categories they belong to, before concluding that the chasm has been bridged in the case of VILC, and to identify in greater detail the forces at work in ensuring uptake of technological innovations by more than a small number of enthusiasts. A longer-term study, similar to that by Egbert, Paulus, and Nakamichi (2002), would provide valuable information on this issue.

Furthermore, individual characteristics are only one of a set of variables that can account for innovation adoption. It may be that reasons for lack of uptake are more complex than the project experience suggests. Dhebar (2001), for example, suggests a further five chasms that need to be bridged before mainstream uptake can be assured. Johnson, Donohue, Atkin, and Johnson (2001) suggest organizational innovativeness is becoming the single most important issue in determining ultimate success of an innovation. Han, Kim, and Srivastava (1998) showed that an organization's innovativeness could affect its business performance. Carlopio (2003) suggests that adopter categories may vary for individuals, who may fall into different categories for different innovations.

However, given the very large investments by governments and funding agencies in online learning, a "bridge-building" component, identifying potential innovations gaps or chasms and the strategies to overcome them, should arguably be part of the implementation plan for every technology innovation in education.

NOTE

1. For more information about these projects and organizations, please see the following websites: aclEnglish, http://www.aclEnglish.com; TAFE Virtual Campus, http://www.tafevc.com.au; and the Virtual ILC, http://www.virtualilc.com.

REFERENCES

Bottomley, Y., Dalton, J., and Corbel C. (1994). *From proficiency to competencies.* Sydney: National Center for English language Teaching and Research.

Carlopio, J. (2003) *Changing gears: The strategic implementation of technology*. New York: Palgrave Macmillan.

Corbel, C. (1996). *The computing practices of language and literacy teachers*. Sydney: National Center for English language Teaching and Research.

Corbel, C. (1998). *Instructional technology strategy for the Adult Migrant English Program*. Sydney: National Center for English language Teaching and Research.

Corbel, C. (1999). *ESL teaching in the global hypermedia environment*. Background Paper No. 4. Australian Council of TESOL Associations.

Corbel, C. (in press). Teacher roles in the global hypermedia environment. In J. Cummins and C. Davison (eds.) *International Handbook of English Language Teaching (Vol. 2)*. Norwell, MA: Springer.

Corbel, C., Gruba, P., and Enright. H. (2002). *Taking the web to task*. Sydney: National Center for English Language Teaching and Research.

Debski, R. and Gruba, P. (1999). A qualitative survey of tertiary instructor attitudes towards project-based CALL. *Computer Assisted Language Learning, 12*(3), 219–39.

Dhebar, A. (2001). Six chasms in need of crossing. *Sloan Management Review, 42*(2), 95–9.

Egbert, J., Paulus, T., and Nakamichi, Y. (2002). The impact of CALL instruction on classroom computer use: A foundation for rethinking technology in teacher education. *Language Learning and Teaching, 6(3)*, 108–26.

Geoghegan, W. H. (1994, July18–20). Whatever happened to instructional technology? [Electronic version]. Paper presented at the 22nd Annual Conference of the International Business Schools Computing Association. Baltimore, MD. Retrieved November 2000, from http://franklin.scale.uiuc.edu/scale/links/library/geoghegan/wpi.html.

Goldsmith, R. E., d'Hauteville, F., and Flynn, L.R. (1998). Theory and measurement of consumer innovativeness: A transnational evaluation. *European Journal of Marketing 32(3/4)*, 340–53.

Han, J. K., Kim, N., and Srivastava, R. K. (1998). Market orientation and organizational performance: Is innovation a missing link? *Journal of Marketing 62(4)*, 30–45.

Interactive Learning Center. (1998, July 15). *Jumping over Moore's chasm?* [Electronic version]. Paper presented at the 1998 Lilly Conference, Anglia Polytechnic University. Retrieved November 2000, from http://www.tsms.soton.ac.uk/ilc/su/pptfiles/conf/lilly/jump.htm.

Johnson, D. J., Donohue, W.A., Atkin, C.K., and Johnson, S. (2001). Communication, involvement, and perceived innovativeness: Tests of a model with two contrasting innovations. *Group and Organisation Management, 26 (1)*, 24–52.

Moore, G. A. (1991). *Crossing the chasm: Marketing and selling technology products to mainstream customers*. New York, NY: Harper Business.

Parks, P., Huot, D., Hamers, J., and H.-Lemonnier, F. (2003). Crossing boundaries: Multimedia technology and pedagogical innovation in a high school class. *Language Learning and Technology, 7(1)*, 28–45.

Rogers, E. M. (1983). *Diffusion of innovations* (3rd ed.). New York, NY: The Free Press.

Tapcott, D. (1999). *Creating value in the network economy.* Boston, MA: Harvard Business Review Press.

5

Developing Information Technology for English in Chinese Secondary Schools

Bob Adamson

INTRODUCTION

Information technology (IT), English, and globalization are interlinked. Advanced communications systems, both hardware and software, have not only become important commodities in the international market; they have also served as powerful media for the forces of globalization. The same holds true for the English language, which is rapidly gaining acceptance as the major international language. But globalization is both benign and destructive. For a country in transition, such as China, technological developments can aid economic modernization; but the transfer of technology (facilitated by proficiency in English language) from other countries, especially those that do not share the same cultural or political values, could have deleterious consequences for the integrity of the state, unless handled judiciously.

This, of course, is not a new problem — the same issues were debated in the 1860s in China, when the Qing dynasty accepted the need to reform the nation in the face of the military and mercantile power of Western countries but sought to do so (not always successfully) without harming its distinct identity. China's response then was to formulate the principle of *zhong xue wei ti xi xue wei yong* (grafting Western practices onto Chinese roots), which meant that foreign practices should be adapted to suit the realities of China (Teng and Fairbank 1979; Shen 1997). The Chinese government set up mechanisms for technological transfer, the study of English being the main strategy, but it controlled the degree of openness of the mechanism (rather like a sluice gate) according to the state priorities of the time. In times of political tension, the technological transfer was tightly controlled to avoid foreign contamination; in a more relaxed climate, the mechanism was loosened: As former paramount leader Deng Xiaoping remarked, if the door is open, it is inevitable that flies will come in. Appendix 1 shows how the sluice-gate mechanism has operated in Chinese education since the Qing dynasty.

This chapter is concerned with technological transfer involving the English language. It describes an IT project involving an American educational software company that was based in Hawaii, Ohana Foundation, and the People's Education Press (PEP), the textbook publishing division of the Ministry of Education in China. The project comprised the development of an innovative IT package to supplement the national *Junior English for China* textbook series that is used in the majority of schools in China. (The package also supplements the *Senior English for China* textbook series, but for reasons of space, the main focus of this chapter is on the junior secondary project.) In particular, the chapter examines how the sensitive issues of transfer were handled by two experienced partners. Over the past fifty years, PEP has evolved a sophisticated process of curriculum development that includes the selective use of foreign approaches to ELT, adapted to ensure that the curricular outcome was pragmatic and endowed with Chinese characteristics (Adamson 1998, 2004); and although Ohana is a new enterprise, senior staff in the foundation had previously worked on many similar cross-cultural projects.

In this study, the Ohana-PEP project is viewed as a form of educational policy. The organization of the chapter is based on four critical dimensions of educational policy identified by Elmore and Sykes (1992): the genesis of the project, the processes of operationalization, the nature of the product, and the impact of the project. This fourth dimension is limited to a discussion of the initial phase of implementation, because the finished software was only being used in classrooms on a limited basis at the time of writing. The main source of data is interviews with key stakeholders; project documents and a review of the product itself provide supporting information. As a consultant to both PEP and Ohana, I had privileged access to the data. Although I have worn two separate hats as a researcher and a contributor to the development of the software, I found that the tension between the two roles was productive. The study has been a form of action research for me, providing a better understanding of the nature and challenges of the project.

THE ENGLISH LANGUAGE CURRICULUM IN CHINA

The PEP-Ohana IT package was designed for a textbook series, *Junior English for China,* linked to the English Language curriculum that was introduced in 1993. The genesis of this curriculum was a number of major educational reforms dating to the mid-eighties, most notably the Nine Years' Provision of Education Law of April 1986 (Adamson 2001). This reform seeks to provide basic education for all children in China — an ambitious undertaking given population growth and the disparities in economic development across the country. A concomitant reform was the decentralization of certain aspects of policy-making in education, such as textbook development, previously the sole preserve of PEP (although the PEP series subsequently dominated the market). Subject-specific reforms included the decision to promote a more communicative approach to English language teaching to serve the new economic profile that had seen a boom in international trade and tourism. However, the lack of trained teachers and the persistence of traditional beliefs concerning the learning styles of Chinese students resulted in the adoption of a cautious, synthetic approach to pedagogical reform that drew upon foreign approaches by adapting them to domestic conditions (Adamson 2001). As a result, the curriculum, as manifested in the published syllabus and the PEP textbook series, had a number of traditional as well as innovative features. The features of the curriculum are summarized in Figure 5.1.

The 1993 syllabus

The 1993 syllabus follows the pattern of its predecessors by setting out the rationale, requirements, and content of the program, as well as the language items to be covered. The learning targets are summarized in Appendix 2. The targets are divided into two levels, to allow for flexibility: Level One is for students to cover in the first two years; Level Two is for the third year (for those areas operating a three-year junior secondary program), or for the third and fourth years (for those areas operating a four-year junior secondary program).

A traditional feature of the syllabus is the control of the level of linguistic difficulty (Adamson 2003). The syllabus quantifies the amount of vocabulary to be taught and the degree of mastery of the various language skills (e.g., for listening, the goal is to achieve seventy percent correctness in understanding after listening three times to native speakers' recording

Genesis	Nine-year compulsory education; strengthening of ELT; national economic modernization; strengthening of the Open Door Policy; decentralization of decision-making
Role and status of English	Economic development: international transfer and communication; access to higher education and lucrative employment; very high official status
Role of agencies	
PEP:	Collaboration in researching and producing the syllabus and textbooks
Foreign publisher:	Co-producer of textbooks; source of technical transfer
ELT specialists:	Consultants in producing the syllabus and textbooks
Teachers:	Consultants; feedback from piloting and post-implementation use
Process	Research in PRC and abroad; collaboration with foreign publisher, specialists, and teachers; extensive piloting; teacher training; continuing revisions in the light of post-implementation feedback
Syllabus	Produced by CTMRI, published by PEP; stresses communicative competence; lists function/notions, as well as vocabulary and grammar
Resources	*Junior English for China*, produced by PEP, plus other textbooks produced by various regions
Junior English for China	
Components	Student's Books, Teacher's Books, Workbooks, Reading Practice Books, cassettes, wall pictures, videos, CD-ROMs, etc.
Organization	Integrated topic-based, cyclical progression with regular revision
Discourse	Anecdotes, stories, everyday dialogues, scientific, and cultural passages
Linguistic components	Focus on realistic oral skills; functions, notions; traditional arrangement of grammatical components
Intended pedagogy	Eclectic: Structural-functional/notional, with some audiolingualism
Political and moral messages	Some moral messages; very weak political content

Figure 5.1 Key features of the English Language Curriculum of 1993

of familiar materials at a reading speed of 100–110 words per minute). Such quantification occurs in most syllabuses published since 1949 and reflects a similar practice in the Chinese Language syllabus. The 1993 syllabus was also innovative in a number of ways. First, it promoted a new, communication-oriented pedagogy:

> It is the first time in the history of EFL in China that the actual use of the language for communication should be placed in such a prominent position. This indicates that schools will no longer teach students about the language but teach them how to use it. (Liu 1995, 2)

This communicative approach is constrained in its authenticity by the strict control of vocabulary and grammar, but it is found in components such as a list of phrases for thirty situations, labeled *Daily Expressions in Communication*. Another innovation is the goal of the syllabus to foster greater understanding of Western culture rather than using English as a propaganda tool against Western imperialism, as was the case in the 1950s and 1960s. At the time of the development of the 1993 curriculum, the political climate was relatively stable, the state stressing economic goals rather than revolutionary activity (unlike, for instance, during the turbulent years of the Cultural Revolution, 1966–76, that was an era of "politics to the fore"). The syllabus states that the curriculum should not be politically overt, but materials should be chosen which, as noted earlier, focus on moral education and foster patriotism and socialism (People's Education Press 1993).

The "Junior English for China" textbooks

Since 1949, there has been an evolution of curriculum development processes and contents for the junior secondary school English curriculum in China (Adamson, 2004). Prior to 1993, the PEP in the Ministry of Education (from 1984–97 known as the State Education Commission) had assumed a central role in curriculum development but had also instituted a system of consultation to allow input from grassroots teachers and specialists in tertiary institutes in China. At the same time, the recommended pedagogy included elements that had become traditional, as well as new elements adapted from overseas (initially from the former USSR and latterly from Western countries).

For the *Junior English for China* series to accompany the syllabus, PEP collaborated with a foreign publisher for the first time, resulting in books that not only are of better technical quality than previous series but also draw on international trends in ELT. To avoid any political pitfalls, PEP excluded political content, preferring, instead, a focus on moral education. The main instructional technique is the *Five Steps* (Revision, Presentation, Controlled Practice, Free Practice and Consolidation), a variation of the Presentation-Practice-Production methodology that was a predominant approach in ELT internationally at the time the books were written. However, despite the involvement of a foreign publisher, it was not direct transfer of Western pedagogy. There was a process of synthesis, the project director, Liu Daoyi, having the final say with regard to the end product. This process is entirely in keeping with the historical maxim of *zhong xue wei ti xi xue wei yong*. As a result, the textbooks incorporate well-established indigenous pedagogical practices, such as the use of memorization, drilling, and the central role of the reading passage. They also follow the sequence of tenses that had been established in textbooks written in the 1950s.

The contents of the textbooks are arranged thematically. The topic of an individual unit serves as the cohesive agent for the linguistic components. Most of Book 1 focuses on oral development, with short dialogues or structure-based spoken drills. This is supported by games, songs, and a range of listening exercises. Some written practice, such as copying and simple compositions, is also found. Reading passages start to appear towards the end of Book 1, and the series gradually becomes more text-oriented.

THE IT INITIATIVE

Genesis

The idea of developing a DVD-based multimedia software package to accompany the *Junior English for China* textbook series originated within the Ohana Foundation, a non-profit organization established in 1999 by Annie Chan, who had enjoyed success in Silicon Valley. Ohana's aim is to introduce cost-effective, innovative technologies in schools, regardless of the economic, geographic, ethnic, or social status of the students. For its international projects, Ohana places emphasis on English:

> Command of the English language is a fundamental requirement for success in the new, digitally based, world economy. Accordingly, Ohana Foundation has dedicated significant resources to the development and delivery of innovative technologies that support English as a Foreign Language (EFL) teaching and learning. (Ohana Foundation 2001)

Chan identified China as the location for Ohana's first major international project. The reasons were complex. First, the demand for English in China is ever increasing, particularly since China's entry into the World Trade Organization and successful bid for the 2008 Olympics (Lam 2002). This, together with the sheer size of the country and the relatively poor conditions in schools in China, meant that there was scope for Ohana to make a significant impact. Furthermore, Chan's own experiences as an ethnic Chinese (and a descendant of the great Confucian teacher, Zhuxi) who had mastered the English language and achieved business success provided a strong motivation for focusing on education in China and ELT as a key factor in the economic development of China. Chan's other ventures had provided her with extensive knowledge in operating in the Chinese context, and key lessons were applied as Ohana sought to approach the relevant agencies in China — business dealings with China tend to require long-term commitment and sensitivity. In this regard, Chan was assisted by senior members of the Ohana staff who shared a similar ethnic and professional background.

One important lesson is the need to deal with the Chinese through the correct channels. Another is the vital role of personal contact. Identifying the Ministry of Education, and PEP as its textbook publishing section, as the conduit for engaging in a major educational enterprise was easy for Ohana: The *Junior English for China* series had been the dominant ELT textbook series in Chinese schools since 1993, used by nearly 40 million students annually, a market share of around seventy percent (Liu 1995). However, establishing a personal contact was not so simple. Ohana's vice president (Marketing) at the time, Lishan Chong, approached a retired professor from the University of Hawaii, Frank Tang (Deng Yanchang), who had taught English for forty years in Beijing. Tang had served as a consultant to PEP on previous curriculum developments at the behest of Liu Daoyi, one of his former students. After looking at Ohana's operations and discussing their plans for China, Tang agreed to provide an introduction to Liu. Ohana invited Liu to Honolulu for initial discussions and later prepared a demonstration of the potential of DVD (with its high-capacity disk space and scope for interactivity) for senior PEP officials in Beijing.

Once the connection with PEP had been established, it was strengthened in various ways. Ohana believed that relevant research should inform the project, so Larry Smith, editor of *World Englishes* journal, was commissioned in conjunction with the East-West Center at the University of Hawaii, to carry out a literature search on the applications of IT in education in general and ELT in particular. This research was linked to a series of workshops on innovation and creativity in ELT held at the East-West Center in 2001, with participants from various Asian countries. Ohana also funded a project led by Liu Daoyi in her capacity as president of the National Foreign Language Teaching and Research Association in China involving six case studies of IT in ELT in Chinese schools. Ohana also promised to undertake a teacher training program to introduce new teaching methods that would incorporate IT and foster innovation and creativity. PEP's confidence in Ohana's *bona fides* was further enhanced when contributors to the original textbook series, Neville Grant, Chris Jacques, Frank Tang and Bob Adamson, and other educators, such as David Eastment and Christina Graf, who were known to PEP, were engaged as consultants.

From PEP's perspective, the Ohana proposal was timely because it matched recent government policy in China that required increased use of multimedia in the delivery of all subjects in the school curriculum, greater attention to fostering creativity in children, and innovation in teaching methods. This required new pedagogical approaches, new teaching materials, and new technology. Also, the project came at a time of decentralization of textbook production, which meant that PEP was facing competition where once it enjoyed a monopoly: The DVD package potentially would make the PEP series more attractive and marketable. Given that Ohana assumed all the development costs, as it would have 100% ownership of the copyright, PEP was positively disposed, as the arrangement excluded PEP from any financial risks. Though cautious, PEP decided to go ahead with the collaboration:

> PEP was at first quite passive. They didn't expect to do such a project because they thought it wouldn't be very profitable, because the DVD is not as popular as the VCD in this country. They thought it would be quite difficult. And without the hardware, schools wouldn't be able to use the software. But PEP thought it was a good idea because with

> the expertise of Americans, they believed the DVD software would be of high quality, and that would support the books. So they thought it was a good idea to co-operate with Ohana. (Liu, Interview, August 10, 2001)

The first stage of the project might be described as a "seduction" model. Ohana identified China as a highly suitable location for its purposes and then instigated a number of strategies to gain the confidence and consent of Chinese counterparts. The latter saw advantages in the partnership, particularly as it engendered little risk, but preferred Ohana to take the initiative.

Processes

Informed by the research projects, marketing analyses, and staff's previous experience, Ohana's approach to designing the *Whiz English* multimedia package was top-down, bottom-up in the sense that it sought to apply state-of-the-art technology in ways that were practicable in the Chinese context. This involved taking into account factors such as the nature of ELT in China as evidenced by the official curriculum and by local practices, as well as the capacity of schools and teachers to adopt and use IT. At the same time, Ohana wanted to enhance education in China with foreign technology while avoiding the pitfalls of cultural, linguistic, or other "imperialisms" that have been identified in similar contexts (Phillipson 1992; Pennycook 1994). Ohana had no illusions about the nature of the challenge, as Andrew Crilly, vice president (Multimedia International), acknowledged: "We wanted to do something with what technology could offer in a learning system that was very, very dated — i.e., just a classroom and a smattering of televisions — but nothing serious for English learning" (Interview, August 2, 2001).

To address the production challenge, Ohana's solution was a three-pronged strategy: (a) using the textbooks as the anchor; (b) supporting and complementing the teacher; and (c) providing technology that was not readily available in China while working closely with Chinese counterparts:

> To make this product, it was always going to be a case of providing an incremental improvement on the learning system in the classroom. So we had to get multimedia that would grow out of the textbooks … It had to be something that was well integrated with what the teacher

> was already doing. This might seem simple, but until you make that kind of decision you don't realise the wisdom of such a step. (Crilly, Interview, August 2, 2001)

The textbooks were therefore identified as the lynchpin of the project. They offered a foundation on which the DVD content could be built and were the mainstay of most teachers' lessons. The practice of using textbooks as the basis for multimedia was derived from the experience of Andrew Crilly and Jane Johnson in making television programs for the Open University in the United Kingdom: "At the Open University, we always knew that you would lose your viewer at home unless you had something meaningful for the message that was in the corresponding text. Otherwise you are wasting your time" (Crilly, Interview, August 2, 2001).

The second aspect, supporting and complementing the teacher, concerned the "value-addedness" of *Whiz English*:

> I took a careful look at the books and I realised … that the textbooks are designed in a visual way because the teachers can't read that well. So therefore we pull everything off the visual elements of those books … adapt them for interactive listening, say, get the digital sound tracks, segment them, because that's what the teacher wants … and you've got a very powerful learning tool. (Crilly, Interview, August 2, 2001)

Adding to the challenge were the self-imposed tight deadlines, designed to ensure the product reached the market as a pioneering innovation. Also, as well as crossing cultural bridges, Ohana needed to blend four disparate areas of expertise: education, production, technology, and marketing. In the event, this was generally a collegial decision-making process, with occasional negotiation and bargaining when compromises were needed.

This is exemplified by the experiences of the project team responsible for developing the content. Ohana carried out a large recruitment exercise, targeting initially production specialists and creative artists, and work started on the first set of *Whiz English* disks. A group of US educationalists were also consulted. The production process started with brainstorming meetings of the staff. As one of the project team leaders, Ed Robello (Interview, July 31, 2001), recalled:

> We would break down the books … and we would have great ideas as to how we could do an animated piece on the disk and really make it fun for the kids … It was not just one or two people; it was the whole team and we were getting their ownership into it.

Work then began on turning the ideas into substance. However, as this work progressed, it became clear — from the team's own perceptions and from feedback received from PEP, the consultants, and teachers, in focus groups set up by Ohana in different parts of China — that more input was needed from the educational sector regarding the Chinese ELT context. Only a few members of Ohana staff had experience in teaching English as a foreign language, and none of them in China. It was at this stage of the project that the consultants were recruited. To enhance in-house expertise, Ohana also requested Liu Daoyi to recommend a secondary school teacher of English from China with a background in computer-assisted language learning. Hu Ming, from Suzhou, arrived in Hawaii in February 2001. His main brief was to view the contents of the DVDs from the perspective of teachers.

Hu's input focused on technical, linguistic, and pedagogical aspects of the software. He cautioned against assuming too much prior IT knowledge on the part of the user: Icons must be readily understood, and the navigational concepts underpinning the interrelationship of components must be familiar. The linguistic complexity of, for example, commentaries for video resources should be controlled in vocabulary and grammar, according to the requirements of the syllabus. He also urged the incorporation of a clear pedagogical application, arguing, "an educational product is different from just an entertaining product" (Hu, Interview, July 27, 2001). Hu was also concerned for the sake of the teachers as end users that, when published, the disks should be a finished product rather than an initial part of an on-going production process: "We mustn't think 'It's just software, so we can have version 1, version 2, version 3 and so on' … for educational products, I don't think we should do it in this way."

Although most of the discussions focused on linguistic, pedagogical, or technical issues, there were, very occasionally, political and cultural aspects that PEP or other Chinese counterparts required Ohana to rectify. For instance, on one frame, the flag of China, although presented as the same size as other national flags, had to be enlarged in the final version, as PEP insisted that the five stars be clearly visible. Likewise, a map of China that inadvertently omitted Taiwan had to be replaced. One cultural problem that arose was the correct pronunciation of the names of Chinese characters that appeared in the DVDs. An accurate rendition of the tones was very difficult for the American and British actors who performed the voice-overs to produce.

In bringing together the disparate perspectives from education, production, technology, and marketing to create a practical product, Ohana adopted a teamwork approach that, according to Ed Robello (Interview, July 31, 2001), had the characteristics of "autonomy, supportiveness, openness, consensus-building … and hard work … and it wasn't just an internal team: We brought people in from the outside and developed a real partnership with PEP."

Team-building in this process was facilitated by the availability of the requisite expertise in relevant areas, which meant that creative solutions were often found to problems. Occasional frustrations occurred, particularly when differences in work culture came into play. For instance, Robello found different approaches to decision-making frustrating: He was used to a production culture in which the resolutions of a meeting were viewed as final, and was disconcerted when others preferred to revisit earlier decisions. Also, the timeframe arising from the marketing strategy caused pressure. It was, as Robello remarked, a difficult learning process at times, but his experience in television broadcasting (which was also the background of other key team members) provided good training for the stresses involved in the production processes:

> I think a lot of frustrations for our team were caused by last-minute changes, but on the other side of the coin, I come out of a television background and we are used to working with horrible deadlines … You have to adjust … I sense that this process has been one big compromise. (Robello, Interview, July 31, 2001)

Overall, the design of *Whiz English* was centered on the challenge of creating high-quality, innovative software for use in a relatively primitive technological environment. The solution adopted was to anchor the DVD content to the textbook series that the end users, the teachers, knew well. The production process often brought about creative solutions achieved through a collegial approach. At times, it also involved compromise, a form of conflict resolution necessitated by the multiple perspectives that were brought by the diverse stakeholders and the pressure of time. However, any frustration engendered by this diversity, by the different work cultures of participants, and by the compromises themselves, was mitigated by the fostering of a shared vision (the focus on the needs of the end user) and of ownership (through team-building).

The nature of the product

In supporting the *Junior English for China* series, the software design identified the contributions that the DVD could make that included the following: dramatized versions of the dialogues in the textbooks, a picture dictionary showing new vocabulary items in each unit, brief documentary segments providing background information on an aspect of the theme of the unit, extension grammar activities, extra listening exercises, and, in the later stages, readings of the passages by storytellers. These elements were to be enhanced by special effects — cartoon clips, interactive multiple-choice questions, subtitles, pop-ups of key vocabulary items, and so on. The rationale behind these choices of components was the desire to use the potential of DVD, with its high-quality visual and audio renditions (including the PAUSE capability that delivers — unlike VCD technology — an enduring freeze frame), in order to support teachers by bringing the outside world into the classroom, showing language in context, and providing a good model of spoken English. Teacher support would take the form of comprehensive training programs and an Activity Guide that would be inserted in the jewel case of the DVD.

In software design, many of these components might appear mundane; but this was precisely the goal of the team — to have any chance of widespread acceptability in China, the resources would have to be within the teachers' zone of proximal development. Also, the components reflect a practical and goal-oriented view of multimedia: The selected medium or media must be the best fitted to serve the designated purpose rather than an elaborate demonstration of all the DVD's "bells and whistles." The design of the software was ultimately conservative: a synthesis of pioneering technology and practical usage in the best traditions of *zhong xue wei ti xi xue wei yong*.

Although the design was conservative, the hardware carried a risk because of the technical innovations it incorporated. The foundation identified interactive DVD technology as a major technical contribution it could make to ELT in China. To enhance the flexibility of the product, Ohana also considered providing Internet-based resources interwoven with those on the disks, using their specially designed interactive DVD-player with a built-in web browser. However, this was both an asset and a potential liability for Ohana. New technology was attractive to potential buyers if the price is right, but the downside was that a supporting infrastructure would need to be created within China. In the initial stages, Ohana worked with selected schools around China, using hardware donated by the foundation, to experiment with the product and to gain a foothold in the market.

Outcomes

The Ohana-PEP project has moved from the drawing board to the delivery stage. Two major issues have been tackled: initial entry into China and the development of the product. The initial feedback from PEP and teachers in the schools experimenting with *Whiz English* was favorable. Liu Daoyi, from PEP, noted, after observing several lessons:

> We could see very clearly that the teachers can teach in a more lively, more communicative way. Some teachers used to just follow the steps as the textbook suggests — they could not make their lessons very lively — but with these materials, because some of the animations are really very good, if you observed the lessons, you would find the students more interested. (Liu, Interview, August 10, 2001)

Feedback gathered directly from the teachers in a questionnaire survey conducted later by Ohana was very complimentary about the range, quality, and flexibility of resources available on the DVD. In particular, the teachers appreciated the authentic environment that they were able to bring to the classroom that enabled the lesson to focus on cultural elements and correct pronunciation. Altogether, seventy-seven percent of the teachers said they would recommend *Whiz English* to their school principal to buy, while twenty-four percent indicated that they would like to buy a personal copy. The only reservation expressed was that some of the contents, despite (or maybe because of) the compromises made during the design process, remained too difficult linguistically or obscure culturally for the students. Overall, however, the indications are that the strategy of linking the DVDs to the textbooks is a sound one.

The technical innovations, inevitably, caused some problems for the teachers. They found the navigation difficult, especially if they were new to IT, and occasionally the loading time was slow. These glitches were being addressed before the finished version of the product was launched. However, the downturn in the IT market after the terrorist attacks on the USA on 11 September 2001 reduced the capacity of Ohana to maximize the potential of the final product, and some technical compromises were necessary. For instance, given the technological challenges and costs associated with the specially designed hardware, Ohana decided not to include the built-in web browser when the product went on general release.

Nonetheless, on the basis of the process of development, the project has broken new ground in crossing international barriers to apply high technology to a low resource environment. It makes a useful contribution to research into innovation and IT through focusing attention on the bilateral nature of the relationship. Figure 5.2 illustrates the interaction that has characterized the Ohana-PEP project. It suggests that effective international transfer — be it technical, pedagogical, or whatever — is not a simple donor-recipient relationship; instead, it is a dynamic, creative process of mutual learning, particular respect being paid to the expertise, practices, and developmental capacity of the end user. Each party has a contribution to make in ensuring that the final product is a practical tool whose applications are immediately apparent to the beneficiaries, in this case, the grassroots students and teachers in China. This notion of effective cultural transfer is in congruence with views of successful curriculum development; a voice must be given to the primary implementers, and consideration must be given to the contexts in which they operate (Morris and Lo 2000; Fullan 2003).

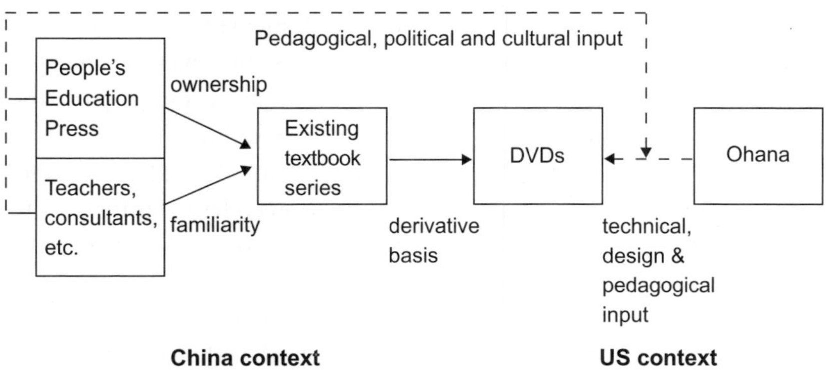

Figure 5.2 Interaction in the design of the Ohana-PEP Whiz English Project

There are still technical, marketing, and pedagogical challenges to overcome. The technological challenge lies in the tensions created by the promotion of pioneering technology in classrooms which have a minimal technical base. Many teachers in China lack the infrastructure and the technical ability to use the resources. They will need support and training, which is a considerable undertaking, given that there are more than a quarter of a million junior secondary school teachers of English in China (and another 150,000 senior secondary school teachers). The marketing challenge involves selling a first-world product to a (largely) third-world country. Distribution and the provision of technical support is difficult in many parts of the country, and pricing of the hardware and software will need to match the budget of schools and individuals, while nonetheless enabling Ohana Foundation to continue to finance the project. The pedagogical challenge remains the need, as the product develops, to continue to temper the creative urges of the software designer with the practical reality of the English language teacher in China. Experience to date suggests that compromises by all sides over a long period of time will be necessary, and, at times, the tensions inherent in Ohana Foundation's commitment to its goals of providing innovative educational technology to all will be sorely tested.

CONCLUSION

The Ohana-PEP project is part of a long history of cultural exchange and technological transfer involving China. Historically, such interaction has not always been a happy or fruitful undertaking. The humiliations of gunboat diplomacy in the nineteenth century raised China's awareness of the deleterious consequences of dealing with inimical trading partners. The relationship established by Ohana was achieved through a strategy of "seduction" that was context specific: an approach through the correct channels and personal recommendations, various relevant initiatives to foster confidence and collaboration, and the anchoring of the product to an existing resource in China, PEP's textbook series. A collegial process, with occasional negotiation and bargaining, was implemented for the development of the product. China's response to the Ohana initiative has historical overtones. Once assured of Ohana's good faith, PEP and teachers in China have worked to ensure that Western technology can be transferred to China in a form that is useful to China's development — the principle of *zhong xue wei ti xi xue wei yong*. The prospects for success of the project was enhanced when Ohana realized the importance of providing scope for Chinese counterparts to play a more active role in tailoring the product to the realities of China; and this reciprocal process will need to continue in tackling the challenges that remain.

ACKNOWLEDGEMENTS

I am very grateful to People's Education Press, Ohana Foundation, and Edward Vickers, for their assistance in the preparation of this chapter.

REFERENCES

Adamson, B. (1998). *English in China: The junior secondary school curriculum 1949–94*. Unpublished PhD thesis, University of Hong Kong.

Adamson, B. (2001). English with Chinese characteristics: China's new curriculum. *Asia Pacific Journal of Education, 21*(2), 19–33.

Adamson, B. (2002). Barbarian as a foreign language: English in China's schools. *World Englishes, 21*(2), 231–43.

Adamson, B. (2003). China's official English: The curriculum for junior secondary schools since 1949. *Humanising Language Teaching* 5(3), major article 1.

Adamson, B. (2004) *China's English: A history of English in Chinese education*. Hong Kong: Hong Kong University Press.

Elmore, R. and Sykes, G. (1992). Curriculum policy. In P. W. Jackson (ed.) *Handbook of research on curriculum* (pp. 185–215). New York: Macmillan.

Fullan, M.G. (2003). *Change forces with a vengeance*. London: Routledge Falmer Press.

Lam, A. (2002). English in education in China: Policy changes and learners' experiences. *World Englishes*, 2(2), 245–56.

Liu Daoyi (1995). English language teaching in schools in China. Paper presented at the International Language in Education Conference, The University of Hong Kong, 13–15 December 1995.

Morris, P. and Lo, M.L. (2000). Shaping the curriculum: Contexts and cultures. *School Leadership and Management* 3(2) 175–88.

Ohana Foundation (2001). China project document (mimeograph).

Pennycook, A. (1994). *The cultural politics of English as an international language*. London: Longman.

People's Education Press. (1993). *Yingyu Jiaoxue Dagang* (English Syllabus). Beijing: People's Education Press.

Phillipson, R. (1992) *Linguistic imperialism*. Oxford: Oxford University Press.

Shen, F. (1997). *Cultural flow between China and outside world throughout history*. Beijing: Foreign Languages Press.

Teng, S. and Fairbank, J.K. (1979). *China's response to the West*. Cambridge, MA: Harvard University Press.

APPENDIX 1

The historical role and status of the English language in China

Period		Role and Status	English Language Education
Later Qing Dynasty	1759–1860	English only permitted to be spoken (in pidgin form) by the despised *compradors* (business intermediaries) in the ports open to foreigners; perceived as a barbaric tongue; low official status	Private study by *compradors* (business intermediaries) in the ports open to foreigners
	1861–1911	Technology transfer: English as a means of access to Western science and technology; helpful to the development of China's international diplomacy; conduit to remunerative jobs in Treaty Ports; later a fad in Shanghai; medium official status	On the curriculum of institutions set up to facilitate technological transfer; later (after 1903) included in curricula of secondary and tertiary institutions
The Republican Era (1911–1949)	1911–23	The Intellectual Revolution: English as a vehicle for exploring Western philosophy and other ideas; opportunities for study abroad; high official status	On the curricula of secondary and tertiary institutions
	1924–49	English for diplomatic, military, and intellectual interaction with the West; resistance from nationalistic scholars and politicians fearing unwanted cultural transfer; medium/high official status	On the curricula of secondary and tertiary institutions
The People's Republic of China (from 1949)	1949–60	Political events make English less favored; used only for gaining access to Western science and technology; low official status	On the curricula of very few secondary and tertiary institutions
	1961–66	The "First Renaissance": popularity of English increases as political pressure wanes; English perceived as valuable for modernization and building international understanding; medium/high official status	Promoted on the curricula of secondary and tertiary institutions

	1966–76	Cultural Revolution: English speakers branded as traitors and punished; Western cultural artefacts attacked; low official status	Removed from the curricula of secondary and tertiary institutions; later restored sporadically
	1976–82	Tentative recovery from Cultural Revolution; English seen as useful for national modernization; medium/high official status	On the curricula of secondary and tertiary institutions
	1982 onwards	English seen as highly desirable for national modernization; important for social, academic and economic success; opportunities for study and travel abroad; especially promoted in key schools; high official status	Strongly promoted on the curricula of secondary and tertiary institutions

Note: From B. Adamson (2002), Barbarian as a foreign language: English in China's schools, *World Englishes*, 21 (2), 233. Reprinted with permission from Blackwell Publishing.

APPENDIX 2

Summary of learning targets for the 1993 curriculum

Level One	Targets
Listening	Understand simple classroom English and give simple answers; understand the main idea of a passage told by the teacher using familiar patterns and vocabulary
Speaking	Carry out simple classroom routines; do question-and-answer exercises in the textbook; use some of the 'Daily Expressions in Communication'
Reading	Understand a reading passage with fewer than 2% new words with a dictionary
Writing	Have neat handwriting and basic orthographic skills; write a dictation based on a familiar passage at a speed of five to seven words per minute; answer questions in the textbook in writing; make simple sentences based on models
Pronunciation	Use the International Phonetic Alphabet and reading rules to pronounce new words; read aloud the reading passage with basic accuracy
Vocabulary	Have an active oral and written command of about 350 common words and 100 common expressions, and passive command of about 300 other words; use phonics
Grammar	Master the inflections and three basic sentence structures; use the present simple, past simple, future simple, and present continuous tenses orally and in writing

Level Two	Targets
Listening	Understand classroom English and give answers; understand the main idea of a passage told by the teacher using familiar patterns and vocabulary; understand with 70% correctness after listening three times to native speakers' recording of familiar materials, at a reading speed of 100–110 words per minute
Speaking	Do simple classroom routines; do question-and-answer exercises in the textbook; use 'Daily Expressions in Communication'; retell passages in simple English
Reading	Read for gist with a dictionary; read aloud and recite coherently; understand a reading passage with fewer than 2% new words with a reading speed of 40–50 words per minute (three-year program) or 50–60 words per minute (four-year program) with 70% correctness
Writing	Have neat handwriting and orthographic skills; write (at around 10–12 words per minute) a dictation based on a familiar passage read at a speed of 100–110 words per minute; answer questions in the textbook in writing; make simple sentences and communicative discourse (such as messages) based on models
Pronunciation	Use the International Phonetic Alphabet and phonics to pronounce new words; read aloud the reading passage accurately; use correct pronunciation and intonation
Vocabulary	Have an active oral and written command of around 600–700 frequently used words plus 200 common expressions; have a passive command of around 400–500 other words; use the rules of reading to remember the spelling of words; tell the meaning and parts of speech of derivatives and compounds; identify the meaning and parts of speech of polysemants in context
Grammar	Master inflections and five basic sentence structures; use present simple, past simple, future simple, present continuous, present perfect, past perfect, and future-in-the-past tenses orally and in writing; master passive voice in present simple and past simple tenses and in sentences containing modal verbs; master usage of infinitives functioning as the object, object complement, and adverbial; understand usage of infinitives as subject or attributive

Note: From *Yingyu Jiaoxue Dagang* (English Syllabus). Beijing: People's Education Press, 1993.

SECTION III

Information Technology as Pedagogy

6 Using Information Technology to Create a Zone of Proximal Development for Academic Language Learning: A Critical Perspective on Trends and Possibilities

Jim Cummins

INTRODUCTION

Russian psychologist Lev Vygotsky (1962, 1978) coined the term *zone of proximal development* (ZPD) to refer to the distance between the level of development or problem-solving that children were capable of on their own, without adult guidance, and the level of development that children could potentially attain under the influence of, or in collaboration with, more capable adults and peers. Expressed simply, the ZPD is the interpersonal space where minds meet and new understandings can arise through collaborative interaction and inquiry. Newman, Griffin, and Cole (1989) label this interpersonal space *the construction zone*, thereby highlighting the considerable influence of Vygotsky's work on the evolution of social constructivist approaches to learning that emphasize the importance of learners' prior experience in the collaborative scaffolding and construction of knowledge.

In this chapter, I consider the extent to which information technology (IT) can provide scaffolds to support the development of first and second language (L1 and L2) literacy abilities, a process I refer to as *academic language learning*. More generally, the chapter examines the relationships between pedagogy and technology. It addresses questions such as: What pedagogical options are available for use of IT in schools? To what extent can IT amplify the impact of different approaches to pedagogy? Is the impact of IT potentially more powerful as a catalyst and support for learning when harnessed to certain forms of pedagogy as compared to others?

These questions reflect the fact that, despite the increase in access to IT in schools, there is still no consensus on *how* computers and other new technologies should be used to support learning. In the United States, for example, schools in affluent middle-class areas increasingly tend to use computers to support higher-order thinking through creative project work, while those in low-income inner city areas still use computers predominantly to reinforce students' mastery of basic skills through drill and practice activities. The periodical *Education Week*, in its *Technology Counts 2001* survey, noted that in most places, the general application of technology with low-achieving students is for "drill and practice" in academic skills. The report notes that, although the "digital divide" between affluent and poor schools has been reduced with respect to quantity indices (schools in high poverty communities have one computer for every 5.3 students, which is just slightly higher than the national average of 4.9), the gap persists in the ways computers are utilized:

> With poor children, student-to-computer ratios belie huge gaps when it comes to Internet access, according to the U.S. Department of Education. In schools where fewer than 11 percent of students qualify for federally subsidized lunches, for example, 74 percent of classrooms have Internet access. On the other hand, in schools where 71 percent or more students are living in poverty, only 39 percent of classrooms are connected to the Internet. (*Education Week*, May 2001, 12)

Furthermore, for low achievers and for students in poor urban schools, technology is typically introduced as a remedial tool involving skills-based software, whereas teachers of more advanced students tend to use a variety of more sophisticated programs (Manzo 2001).

The overall impact of IT on student achievement also remains unclear. Despite small-scale studies suggesting that IT can boost reading comprehension (National Reading Panel, 2000), there is little evidence of any overall impact on national reading levels as a result of more than a decade of increasing access to IT in US schools. Bennett (2002), for example, notes that between 1994 and 1999, there was a massive infusion of technology in US schools, but overall improvements in educational achievement as measured by the National Assessment of Educational Progress were minimal:

> Results for 1999 showed no significant change in reading, mathematics, or science for the three age groups tested—9-year-olds, 13-year-olds,

> and 17-year-olds—from 1994 through 1999. During this five-year period, schools acquired huge numbers of computers and hoped earnestly that this influx of technology would improve education. (p. 622)

Not only has there been no improvement during the 1990s in reading scores, the gap between the highest- and lowest-performing students on the tests has increased. Between 1992 and 2000, the average performance of the top twenty-five percent of students improved while the bottom twenty-five percent declined (Barton 2001; National Assessment of Educational Progress 2001). Barton (2001) also reports that between 1992 and 1998, the gap between white and minority (African American and Hispanic) students in fourth grade reading scores increased significantly in six states and decreased in only one.

An implication of these trends is that IT must boost the achievement of currently underachieving low-income and marginalized students if it is to exert any meaningful impact on achievement levels generally. Thus, these students are a major focus of the present discussion of the potential of IT to promote academic language learning.

The lack of evidence for the effectiveness (or cost-effectiveness) of expanding IT provision has not lessened the zeal of most governments in industrialized countries, supported by business interests and parents, to increase the quantity and quality of computer provision in schools. The assumption appears to be that IT is an integral part of the new economy of the twenty-first century, and IT should therefore be an integral part of the school systems that are supposed to prepare students for this new economy.

I argue that the potential of IT to boost academic achievement has remained largely unrealized because the focus of provision has been on technology rather than on pedagogy. Without a coherent *pedagogical* framework for conceptualizing language and literacy development in schools, the potential impact of IT as an amplifier of pedagogical intervention and student achievement cannot be assessed. In discussing the pedagogical possibilities of IT, I distinguish among three pedagogical orientations: *traditional, social constructivist,* and *transformative.* I provide examples of IT applications within each of these three orientations and suggest that the potential of IT is most clearly evident when schools adopt a transformative pedagogical orientation. I then describe a transformative pedagogical framework for promoting academic language learning (Cummins 2001) and show how such a framework can generate more coherent policy and practice with respect to IT in education. The framework is used to interpret case study data deriving from computer-mediated sister class networks (Brown, Cummins, Figueroa, and Sayers 1998; Cummins and Sayers 1995; Kourtis-Kazoullis and Skourtou, in press).

ORIENTATIONS TO PEDAGOGY

Overview

The three orientations to pedagogy differ in the extent to which the teacher retains exclusive control over classroom interaction as opposed to sharing some of this control with students. The dominant instructional model in most Western industrial societies has been termed a *banking* model (Freire 1970; Freire and Macedo 1987) on the grounds that teachers are expected to deposit information and skills in students' memory banks. This traditional model can be contrasted with social constructivist and transformative models of pedagogy. Social constructivist pedagogy historically is associated with the progressive pedagogy of John Dewey (1963), who emphasized the centrality of student experience and the importance of encouraging active student learning rather than passive reception of information. Although a variety of labels has been used, I am using the term *social constructivist* as an umbrella term to refer to the pedagogical assumptions that are associated not only with Dewey's work but also the work of socio-cultural theorists whose pedagogical focus derives primarily from the work of Vygotsky (1962, 1978) (e.g., Tharp and Gallimore 1988; Wells 1999).

In philosophical debates about education throughout this century, traditional and constructivist orientations have vied for ascendancy at regular intervals. The focus on the transformative potential of education is a more recent phenomenon and is strongly influenced by Paulo Freire's work. Each of the three orientations incorporates a set of instructional and social assumptions. Instructional assumptions are concerned with the conceptions of language, knowledge, and learning that underlie various forms of teaching; social assumptions focus on the ways in which relations of culture and power are addressed in the curriculum.

Most proponents of traditional and constructivist pedagogies tend to focus more on instructional than on social dimensions. They tend to see their instructional recommendations as socially neutral and non-ideological. By contrast, advocates of transformative pedagogy argue that all forms of instruction entail social assumptions, whether acknowledged explicitly or not. The forms of thinking and literacy that are encouraged in school anticipate the forms of civic participation that students are being prepared to undertake upon graduation. Transformative pedagogy explicitly aims to prepare students to participate fully in shaping their societies and to uphold principles of human rights and social justice that are enshrined in the constitutions of most Western industrialized countries.

Although these three orientations are expressed as distinct categories, it is more appropriate to see them as points on a continuum that merge into one another. For example, a transformative orientation will usually include considerable explicit instruction, and much classroom interaction will be constructivist in nature rather than focused directly on social realities. However, a transformative orientation will also actively seek opportunities to relate instruction to students' experience and identity and to the social realities within which their experiences and identities are embedded. Each of these orientations is considered in more detail in the sections that follow.

Traditional pedagogy

The focus of traditional pedagogy is on transmission of information and skills. In the case of L2 teaching, this implies teaching language structures and forms with little emphasis on internalization of meaning or active communicative/authentic use of the language. Traditional pedagogy is exemplified when students learn predominantly from the textbook or when instruction follows fully scripted lessons. In defending the use of instructional scripts, Becker (1977) provides a clear statement of the behaviorist foundations of this form of instruction:

> The use of explicitly detailed lessons — scripts — has been criticized as restricting teachers' initiative. This may be a valid criticism, but one should consider the potential advantages of scripts in providing quality control in a delivery system. The scripts permit the selection and testing of sequences of examples that produce efficient learning if followed. Most teachers simply do not have time to find appropriate words and examples or to sequence skill hierarchies in the most efficient possible manner. When teachers phrase their own questions, they may choose terms unknown to lower-performing children or may include unnecessary verbiage. In choosing examples, moreover, they may teach

incorrect rules because the positive examples have some irrelevant feature in common. (p. 523)

Within this orientation to teaching language or reading, IT is typically used to reinforce the learning of phonics, vocabulary, and grammatical knowledge. The format is frequently made more appealing to students than the textbook format by means of interactive games and activities, but the basic aim is to reinforce information and skills.

Social constructivist pedagogy

Within social constructivist approaches, students are encouraged to construct meanings and become cognitively engaged in challenging projects and activities. Students' prior knowledge is systematically activated, and they are encouraged to bring their experience or *cognitive schemata* to bear on solving problems. During the past twenty years, much of the research within this social constructivist tradition has drawn on socio-cultural theory, most notably Vygotsky's notion of the *zone of proximal development*. An example of the use of IT within this social constructivist orientation is provided by Bracey (2000), in discussing a project in which her students worked with scientists from the National Aeronautics and Space Administration (NASA):

> In one NASA project I did with my 4th and 5th graders called Marsville, we connected with other classrooms online to design a Marsville City (www.challengercenter.org/tr/tr_prpro_set.htm). Marsville was a project-based activity where students created a prototype habitat for Mars. The children came together to learn and build their city and make their own living spaces using a variety of interdisciplinary skills. In the process, they learned creative problem-solving, cooperative learning and data analysis. We studied the systems needed to survive on Mars. We did not just read about it, we did it! It was exciting to see how such a project engaged my immigrant students who were still learning English, and motivated their reading and science learning. (pp. 4–5)

This example clearly involves higher-order thinking skills and cognitive challenge. It also is likely to be much more highly motivating for students than simply learning about Mars or human habitats from a textbook. The active, hands-on, co-operative activities also appeared to be cognitively engaging and motivating for students who were learning through an L2. In this case, students are learning both scientific content and L2 structures and functions simultaneously.

Transformative pedagogy

The instructional assumptions of transformative pedagogy are similar to those of social constructivist pedagogy. However, they diverge with respect to social assumptions. Transformative pedagogy uses collaborative critical inquiry to enable students to analyze and understand the social realities of their own lives and of their communities. Students discuss, and frequently act on, ways in which these realities might be transformed through various forms of social action. Thus, the focus is on constructing meanings, cognitive challenge, and dynamic support within the zone of proximal development. However, the *content* for investigation and inquiry has *social relevance* related to the power structures in society. For example, within a transformative pedagogical orientation, the Marsville project described by Bracey (2000) might be followed up by a project that examines the problems of human habitats *on earth* and the causes of these problems. Students might design an urban habitat on earth that avoids current problems (e.g., homelessness, violence, poverty, pollution, etc.). This would require that they research and analyze, for instance, sources of inequity in income, causes of pollution, and discuss how such problems can be resolved. These forms of inquiry focus directly on and potentially challenge coercive relations of power in the society at large. This pedagogical challenge to societal power relations is particularly significant for marginalized students who frequently come from communities that have been on the receiving end of coercive relations of power over generations.

A variety of case studies of computer-mediated sister-class networks that employ a transformative orientation have been described by Brown, Cummins, Figueroa, and Sayers (1998); Cummins and Sayers (1995); and De Klerk (1998). However, the use of IT within this orientation is still relatively infrequent compared to its use within traditional or constructivist orientations. This is not surprising, in view of the fact that issues related to societal power relations are seldom included in school curricula.

The next section considers in more detail the nature of academic language proficiency, and then a pedagogical model derived from research on academic language learning in both first and second languages is elaborated. This framework represents a starting-point for considering the pedagogical underpinnings of technology-supported academic language learning.

WHAT IS ACADEMIC LANGUAGE PROFICIENCY?

Academic language proficiency can be distinguished from two other dimensions of language proficiency: *conversational fluency* and *discrete language skills* (Cummins 2001). The rationale for making these distinctions is that each dimension of proficiency follows very different developmental paths among both L2 learners and native-speaking students, and each responds differently to particular kinds of instructional practices in school.

Conversational fluency represents the ability to carry on a conversation in familiar face-to-face situations. The vast majority of native speakers of English have developed conversational fluency when they enter school at age five. This fluency involves use of high-frequency words and simple grammatical constructions. English-as-a-second-language (ESL) students generally develop fluency in conversational aspects of English within a year or two of intensive exposure to the language, either in school or in the environment.

Discrete language skills reflect specific phonological, literacy, and grammatical knowledge that students can acquire in two ways: (a) through direct instruction; or (b) through immersion in a literacy- and language-rich environment, either in the home or in school. Students exposed to a literacy-rich environment in the home generally acquire initial literacy-related skills, such as phonemic awareness and letter-sound correspondences, with minimal difficulty in the early grades of schooling.

Some of these discrete language skills are acquired early in schooling and some continue to be acquired throughout schooling (e.g. spelling). The discrete language skills acquired early include knowledge of the letters of the alphabet, the sounds represented by individual letters and combinations of letters, and the ability to decode written words into appropriate sounds. Some of these skills, such as phonemic awareness and knowledge of the letters of the alphabet, show consistently moderate relationships with the acquisition of word decoding skills (Snow, Burns, and Griffin 1998; National Reading Panel 2000).

ESL students can learn these specific language skills concurrently with their development of basic vocabulary and conversational fluency. However, little direct transference is observed to other aspects of oral language proficiency, such as linguistic concepts, vocabulary, sentence memory, and word memory (Geva 2000; Kwan and Willows 1998).

Academic language proficiency includes knowledge of the less frequent vocabulary of English as well as the ability to interpret and produce increasingly complex written language. As students progress through the grades, they encounter far more low-frequency words (primarily from Greek and Latin sources), complex syntax (e.g. passives), and abstract expressions that are virtually never heard in everyday conversation. Students are required to understand linguistically and conceptually demanding texts in the content areas (e.g., literature, social studies, science, mathematics) and to use this language in an accurate and coherent way in their own writing.

The linguistic demands of the curriculum become more acute around the grade four level. Chall, Jacobs, and Baldwin (1990) point out that at grade four and beyond, "the reading materials become more complex, technical and abstract and are beyond the everyday experiences of most children" (p. 45). By contrast, in the earlier grades, vocabulary and reading tests assess students' knowledge of common, familiar, and concrete words.

> Whereas the major hurdles prior to grade 4 are learning to recognize in print the thousands of words whose meanings are already known and reading these fluently in connected texts with comprehension, the hurdle of grade 4 and beyond is coping with increasingly complex language and thought. (Chall, Jacobs, and Baldwin, p. 45)

Acquiring academic language is challenging for all students. For example, schools spend at least twelve years trying to extend the conversational language that native-speaking children bring to school into these more complex academic language spheres. It is hardly surprising, therefore, that research has repeatedly shown that ESL students, on average, require *at least* five years of exposure to academic English to catch up to native-speaker norms (Cummins 1981; Klesmer 1994).

In short, the distinctions outlined above reflect the fact that many ESL students who have acquired conversational fluency and decoding skills in English are still a long way from grade-level performance in academic language proficiency (Cummins 2001; Krashen 2001). Students who can "read" English fluently may have only a very limited understanding of the words they can decode. Reading comprehension does not develop automatically on the basis of decoding skills; similarly, academic language proficiency is largely unrelated to students' conversational fluency in English or their knowledge of discrete language skills. A core component of academic language proficiency is vocabulary knowledge, and the vocabulary load in the curriculum increases dramatically after the primary grades. The development of academic language proficiency, for both ESL and non-ESL students, requires that students gain access to academic language by means of extensive reading and classroom discussion, and that they are supported in harvesting the language they encounter in literature and content area texts.

Thus, two criteria for assessing any computer program or computer-mediated project that aspires to promote academic language proficiency is that (a) it engage students in extensive reading and (b) provide supports for understanding textual meaning and the learning of vocabulary, grammar, and discoursal features of the target language that go beyond what is typically available in reading hard-copy text. Additionally, the content of any reading or inquiry that students undertake should be of sufficient interest and relevance to them that it generates what I label *identity investment*. In the absence of identity investment, cognitive and academic engagement is unlikely to be sustained.

In the following section, I elaborate a pedagogical framework that may serve as a starting point for discussion of how IT might be mobilized to promote academic language proficiency. This framework is consistent with social constructivist approaches to pedagogy but goes beyond constructivist approaches insofar as it emphasizes the importance of encouraging students to develop a critical awareness of how language and power intersect and to use language to explore and act on social realities.

A FRAMEWORK FOR TECHNOLOGY-SUPPORTED ACADEMIC LANGUAGE LEARNING

The central sphere in Figure 6.1 represents the interpersonal space created in the interactions between teachers and students. Within this interpersonal space, knowledge is constructed and identities are negotiated. In contexts of cultural, linguistic, or economic diversity, these interactions are never neutral: they either challenge the operation of coercive relations of power in the wider society, or they reinforce these power relations. Within the interpersonal space of teacher-student interactions, the more students become engaged cognitively, the more academic progress they are likely to make. Engagement of higher level cognitive processes is likely to produce much more learning, for the simple reason that much more of students' brains are involved in the process of learning. If students are involved primarily in rote memorization in the classroom, only a fraction of their cognitive capacity is engaged in learning.

The interpersonal space represented in Figure 6.1 extends the notion of the *zone of proximal development* beyond the cognitive sphere into the realms of affective development and power relationships. The dual process of reciprocal negotiation of identity and collaborative generation of knowledge take place within this *construction zone* and are seen as being intimately related to each other. Teacher-student collaboration in the construction of knowledge will operate effectively only in contexts in which students' identities are being affirmed. The framework also makes clear that the *construction zone* can also be a *constriction zone* where student identities and learning are constricted rather than extended.

The relevance of identity investment derives from the fact that affect is a major determinant of the extent to which students are likely to engage cognitively and academically. Students will be reluctant to invest their identities in the learning process if they feel their teachers do not like them, respect them, and appreciate their experiences and talents. In the past, students from marginalized social groups have seldom felt this sense of affirmation and respect for language and culture from their teachers, and consequently their intellectual and personal talents rarely found expression in the classroom. This perspective implies that in considering the potential of IT projects or programs to increase academic engagement among students from marginalized communities, two significant factors are likely to be (a) the extent to which students are encouraged to explore and appreciate aspects of their own cultural and linguistic heritage and (b) the extent to which they come to see themselves as intelligent and capable human beings in the process of carrying out these projects.

There is a reciprocal relationship between cognitive engagement and identity investment. The more affirmed and respected students feel in the learning environment, the more they are likely to engage cognitively; cognitive engagement results in learning, and the more students learn, the more their academic self-concept grows, which, in turn, results in more cognitive and academic engagement.

The close linkage between identity investment and cognitive engagement can be illustrated with respect to the role of students' prior knowledge in the teaching/learning process. There is general agreement among cognitive psychologists that we learn by integrating new input into our existing cognitive structures or schemata. Our prior experience provides the foundation for interpreting new information. No learner is a blank slate. In reading, for example, we construct meaning by bringing our prior knowledge of language and of the world to the text. As Fielding and Pearson (1994) point out, research conducted in the late 1970s and early 1980s consistently revealed a strong reciprocal relationship between prior knowledge and reading comprehension ability: "The more one already knows, the more one comprehends; and the more one comprehends, the more one learns new knowledge to enable comprehension of an even greater array of topics and texts" (p. 62).

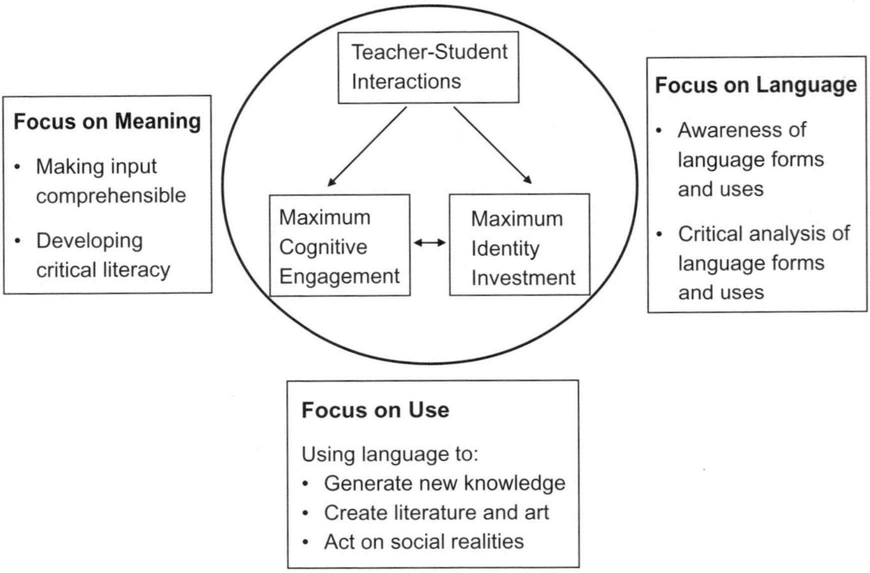

Figure 6.1 The development of academic expertise

Thus, in making input comprehensible, particularly to L2 learners, it is important to activate students' prior knowledge and, where necessary, build background knowledge. This increases students' cognitive engagement and enables them to function at an intellectually and linguistically higher level. Students understand more and, consequently, they learn more language and academic content.

However, just as important for the learning process as these cognitive considerations is the fact that activation of prior knowledge enables teachers to validate culturally diverse students' background experiences and affirm their cultural knowledge. Inviting students to contribute what they already know to the class discussion communicates to students that the cultural and linguistic knowledge they are bringing into the classroom is important and that their voice will be heard in this learning community. Identity is being negotiated in ways that motivate students to express their growing sense of self and participate actively in the learning process.

With specific reference to academic language learning, the framework proposes that instruction should focus on *meaning, language,* and *use*. Each of these foci is sketched below.

Focus on meaning

The framework highlights the fact that effective instruction in an L1 or L2 must focus initially on meaning or messages. Virtually all applied linguists agree that access to sufficient comprehensible input in the target language is a necessary condition for language acquisition.

The Focus on Meaning component argues that the interpretation of the construct of *comprehensible input* must go beyond just literal comprehension. Depth of understanding of concepts and vocabulary as well as critical literacy are intrinsic to the notion of comprehensible input when we are talking about the development of academic language proficiency. This implies a process whereby students relate textual and instructional meanings to their own experience and prior knowledge (i.e., activate their cognitive schemata), critically analyze the information in the text (e.g., evaluate the validity of various arguments or propositions), and use the results of their discussions and analyses in some concrete, intrinsically motivating activity or project (e.g., making a video or writing a poem or essay on a particular topic).

Brown (2000) provides an example of how students as young as grade two are capable of collaborative critical inquiry within the context of an international telecommunications project. The project was entitled *The Eradication of Poverty* and involved classes in different parts of the world exploring the origins of poverty and how to reduce it. According to Brown:

> Second graders in California's Central Valley whose parents worked long days in the fields designed and shared a survey that investigated the assumption that if one just tried harder, one wouldn't be poor. By interviewing people of lower-, middle-, and upper-class occupations in their town, students came to question this assumption. They began to see the links between poverty, race, and educational opportunities. (p. 317)

In short, for learning of academic content, the notion of *comprehensible input* must move beyond literal, surface-level comprehension to a deeper level of cognitive and linguistic processing. This process was evident in the *Marsville* project considered earlier, in which students' engagement with cognitively challenging problem-solving enabled them to move to a much deeper level of understanding than would likely have been the case with learning similar content from a textbook. Equally challenging from a cognitive point of view was the *Eradication of Poverty* project described by Brown (2000), but unlike the *Marsville* project, it brought issues of habitat and life opportunities back home to earth and to students' own communities. Students came to understand more of the social realities of their own communities and were enabled through telecommunications to share and discuss their observations and insights with geographically distant collaborators. The sister-class network expanded the interpersonal space significantly beyond what is usually possible in a single classroom.

Depth of cognitive engagement seldom occurs within a traditional pedagogical orientation, because the focus is typically on memorization of content and internalization of particular literacy-related skills. Thus, the comprehensible input that students receive seldom moves beyond superficial literal understanding.

Focus on language

The Focus on Language component in Figure 6.1 attempts to put controversial issues such as the appropriate time and ways to teach grammar, the role of phonics in reading instruction, etc. under the "umbrella" of *Language Awareness*. The development of language awareness includes not just a focus on formal aspects of the language but also the development of *critical language awareness* that encompasses exploration of the relationships between language and power. Students, for example, might carry out research on the status of different varieties of language (e.g., colloquial language versus formal "standard" language) and explore critically why one form is considered by many educators and the general public to be "better" than the other. They might also research issues such as linguistic code-switching and the functions it plays within their own lives and in their bilingual communities. Or, they might analyze letters to the editor on controversial issues such as immigration and examine how the language used in these letters positions and potentially stereotypes minority group learners such as them and their parents.

In short, the framework proposes that a focus on formal features of the target language should be integrated with critical inquiry into issues of language and power. Also, to be effective, a focus on language must be linked to extensive input in the target language (e.g., through reading) and extensive opportunities for written and oral use of the language.

A number of scholars and educators have focused on the importance of developing language awareness not only as a means of demystifying language and how it works but also as a way of reinforcing students' sense of identity. Lisa Delpit (1998), for example, talks about encouraging African American speakers of Ebonics (i.e., African American varieties of English) to become "language detectives" investigating similarities and differences between their own vernacular and other forms of English such as that found in school texts. For example, groups of students can work together to create bilingual dictionaries of their own language forms and Standard English. A significant goal is to reinforce students' understanding that their language is legitimate and powerful in its context of use but that other forms of English are necessary in different contexts of use.

Figure 6.2 outlines some of the activities that constitute a Focus on Language. These activities clearly go beyond just the teaching of forms and functions of the language. The goal is to develop among students a culture of inquiry into language and how it works in different social situations.

- The structure of language systems (e.g., relationships between sounds and spelling, regional and class-based accents, grammar, vocabulary, etc.);
- Ways of accomplishing different functions and purposes of language;
- Conventions of different musical and literary forms (e.g., rap, rock, folk music, poetry, fiction, etc.);
- Appropriateness of expression in different contexts (cultural conventions of politeness, street language versus school language, the language of everyday speech versus the language of books, language variety as a badge of identity in groups as diverse as gangs, political parties, fraternities, etc.);
- Ways of organizing oral or written discourse to create powerful or persuasive messages (e.g., oratorical speeches, influential written documents, political rhetoric, advertisements, etc.);
- Cross-lingual comparison of languages (e.g., cognates between English and Greek, proverbs, orthography, etc.)
- Diversity of language use in both monolingual and multilingual contexts (code-switching in bilingual communities), language maintenance and loss in families, political controversies surrounding language.

Figure 6.2 Collaborative inquiry to develop critical language awareness

Computer technology can be useful in helping students (either individually or in groups) to collect, internalize, and consolidate their knowledge of language and then use it powerfully to extend their intellectual horizons and personal identities. For example, extrapolating from the paper-and-pencil activities suggested by Norah McWilliam's (1998) *word-weaving* project, students could set up templates in computer files to enter words that they have come across in their reading or everyday experiences that they want to explore. The templates might include categories such as *synonyms, L1 equivalents, proverbs, idioms, and jokes in which the word appears, advertisements, puns, relevant grammatical information,* etc.

IT provides a wealth of both people resources and information resources for students and educators working to develop critical language awareness. One example is the *Orillas* International Proverbs Project described by Brown (1998). *Orillas* is a multilingual sister-class project that has operated for twenty years to promote intercultural learning and academic success for linguistically diverse students (see Brown, Cummins, Figueroa, and Sayers 1998). The International Proverbs Project used telecommunications to enable students from Spanish-, Portuguese-, and English-speaking communities to collect, share, and critically analyze proverbs. Brown (1998) describes one classroom as follows:

> Ms Hernandez's class, like others in the Orillas network, gathered proverbs from family, friends and neighbors. Little by little the list grew. "Nearly a hundred animal proverbs, alone!" the class exclaimed.
>
> Sorting the proverbs wasn't easy. The students debated the meanings of the proverbs, identified different versions and regional variations and compared notes about the contexts in which their parents and grandparents used the proverbs. As they gathered and analyzed, they stayed in touch with other classes on the network. (p. 15)

Parent involvement was dramatically increased in some contexts, as a result of the project. For example, in Watsonville, California, a migrant farmer and his kindergarten child wrote the following critique of the proverb *A bird in the hand is worth two in the bush*: "We don't agree with this proverb because people shouldn't be satisfied with what we have but instead should struggle and make an effort to make each day better" (Brown, Cummins, Figueroa, and Sayers 1998, 15).

Teachers in the project felt that this focus on proverbs represented a powerful way of building up students' critical awareness of language and culture. Among the advantages of focusing on proverbs are the following:

1. Proverbs are virtually universal, permitting cross-cultural comparison.
2. Families become involved, oral histories are encouraged, and students can take pride in the richness of their heritage, as represented in the proverbs.
3. Analyzing proverbs encourages discussion, categorization by themes, critical thinking, and an appreciation of culture.
4. Young students can participate, since the actual amount of text is small and easily typed.
5. Proverbs can encourage much longer writing, such as opinion statements or modern fables.

The International Proverbs Project illustrates once again how telecommunications, implemented within a transformative pedagogical orientation, can expand the interpersonal space or zone of proximal development. Teacher-student interactions have expanded to include family-student interactions as well as peer interactions, both locally and globally. Although the immediate focus is on a language phenomenon (proverbs), both cognitive engagement and identity investment are increased as a result of the fact that the students themselves have collected the proverbs and the proverbs represent their cultural heritage. Thus, students have ownership of the project and of its outcomes. In the context of the project, knowledge has been generated and identities affirmed in ways not possible within a traditional curriculum.

Focus on use

The Focus on Use component is based on the notion that L1 and L2 acquisition will remain abstract and classroom-bound unless students have the opportunity to express themselves — their identities and their intelligence — through language. In order to motivate language use there should ideally be an authentic audience that encourages two-way communication in both oral and written modes. Clearly, computer-mediated sister-class projects provide such an audience. The three examples of language use presented in Figure 6.1 (*generate new knowledge, create literature and art, act on social realities*) are intended to illustrate important components of critical literacy. Language must be used to amplify students' intellectual, aesthetic, and social identities if it is to contribute to student empowerment, understood as the *collaborative creation of power*. Unless active and authentic language use for these purposes is promoted in the classroom, students' grasp of academic (and conversational) aspects of their L2 is likely to remain shallow and passive.

There is little question that IT can provide many of the essential components required to stimulate active written language use. Collaborative sister-class projects, publication of student work on classroom or school webpages, or simply the use of computers to lay out and print newsletters or other forms of publication all facilitate access to wider audiences than would otherwise be possible. IT can dramatically expand the communities of inquiry to which students have access and provide immediate outlets for communicating the results of students' intellectual and artistic work (e.g., through school or class webpages).

Brown (2000), for example, describes a project dubbed "New Places" by the participating classes in which students who had moved to new communities described their experiences. Students who hadn't moved interviewed peers at their school about how they were received in their new schools and communities. Students from a dozen countries were involved in this project, including those who had moved from rural China to Beijing, African Americans who had moved from the south to the north within the United States, and immigrant students who had moved from many countries to the United States. Students investigated what motivates migration and how people from different cultural, racial, and linguistic backgrounds were received by their new communities. Brown concludes:

> One of the outcomes of this project was that together the students analyzed the linguistic, cultural, and institutional barriers at their schools and drew up guidelines for teachers and students about how to make their schools better places for newcomers. The idea that collaborative problem solving might make the world a better place motivates much online learning. (Brown 2000, 312)

As in many examples of sister-class collaboration, the power of a project such as this derives from its dual and complementary focus on knowledge generation and identity negotiation. Clearly, depending on the sophistication of the equipment available to learners, video and audio components can be integrated into such projects.

CONCLUSION

A framework for computer-supported academic language learning has been outlined that views the interactions between educators and students as the most immediate determinant of student success or failure in school. These interactions can be viewed through two lenses: the lens of *the teaching-learning relationship* in a narrow sense, represented by the strategies and techniques that teachers use to provide comprehensible input and reading instruction as well as promote content knowledge and cognitive growth; and the lens of *identity negotiation* that is represented by the messages communicated to students regarding their identities — who they are in the teacher's eyes and who they are capable of becoming.

Instruction can increase cognitive engagement and identity investment when the power of IT is harnessed to connect students with sister-classes in different cultural, linguistic, and geographical contexts. Students' own identities can be expanded when they have to define aspects of their own reality to present to the sister-class and when this reality is reflected to them through discussion and project collaboration. Literacy is expanded in the process of carrying out joint projects, for the simple reason that students read much more and write much more than when they are in an isolated, contained, non-communicative classroom situation.

So why, then, are we not seeing an explosion of sister-class projects that maximize both cognitive engagement and identity investment? In North America, certainly, the technology is there; but, as already noted, for minority and low-income students, computers are typically used for low-level drill and practice routines (traditional pedagogy) rather than for creative and challenging collaborative work (constructivist pedagogy). Projects rooted in a transformative pedagogy orientation are still extremely rare, despite their greater potential to promote cognitive engagement, knowledge generation, and identity investment.

I believe the reasons have a lot to do with control and fear: the feeling among policy-makers and many parents that we need to exercise tight control over the curriculum so that what is taught will determine what can be thought. The other side of this coin is fear: fear that interacting with and getting to know *the other* will undermine our in-group identification and the transmission of *our* values unchanged to *our* children. Most societies still remain cautious and sometimes even xenophobic in relation to those from other cultures, "races," and religions.

Currently, as educators, we have the instructional and technological tools to amplify students' intellectual, linguistic, and moral development more powerfully than at any time in human history. The cognitive engagement and identity investment that is stimulated by sister-class projects involving technology-supported collaborative critical inquiry is immense. But for this to happen, we have to build and continually refine explicit school policies that focus simultaneously on language, technology, and intercultural education. These policies provide maps that help us figure out how to get where we want to go. However, the real challenge is to figure out how much critical literacy and critical thinking we want our students to develop. Critical literacy is a double-edged sword that, in any society, can be focused on our own inequitable social realities, myths of nationhood, historical omissions, religious beliefs, and cultural myopia as much as on "safe" technical problems. It is much safer to remain within a traditional pedagogical orientation in which pre-scripted content is transmitted for students to internalize without elaboration or dissent than to conceptualize the zone of proximal development as a knowledge construction zone where social understandings can be negotiated in the process of critically analyzing academic content. The significant increase in standardized test use in North America and elsewhere in recent years reflects this determination to confine students within the cocoon of the sanitized curriculum. Clearly, adequate evaluation of the technology-mediated student work discussed here would require that standardized assessment be at least augmented to include portfolio assessment.

In summary, the fact that massive investment in computer technology has failed, thus far, to yield appreciable gains in student achievement has very little to do with technology itself. Student failure in most contexts is rooted in social inequities and the associated pattern of historical and current power relations. As reflected in the digital divide" phenomenon, the introduction of IT into schools has tended to reinforce rather than challenge these coercive power relations, at least in North America. These realities, however, should not obscure the fact that IT does have enormous potential to promote student achievement when it is harnessed to appropriate forms of pedagogy. The examples outlined above, together with the pedagogical framework proposed in the present chapter, suggest that IT represents a highly effective tool both to challenge coercive relations of power and promote academic language learning among students from marginalized communities, when it is rooted in a transformative pedagogical orientation.

REFERENCES

Barton, P. E. (March, 2001). *Raising achievement and reducing gaps: Reporting progress toward goals for academic achievement.* A report to the National Education Goals Panel. Washington, DC. Retrieved June 24, 2002, from http://www.negp.gov.

Becker, W. C. (1977). Teaching reading and language to the disadvantaged. *Harvard Educational Review, 47*(4), 518–43.

Bennett, F. (2002). The future of computer technology in K-12 education. *Phi Delta Kappan, 83*(8), 621–5.

Bracey, B. (Spring, 2000). A different divide: Teachers and other professionals. *Edutopia,* 4–5. Retrieved from http://www.glef.org.

Brown, K. (1998). Global learning networks. In G. De Klerk (ed.) *Virtual power: Technology, education, and community* (pp. 8–17). Long Beach, CA: Pacific Southwest Regional Technology in Education Consortium.

Brown, K. (2000). Global learning networks: Heartbeats on the Internet. In J. Vilamil Tinajero and R. A. DeVillar (eds.) *The power of two languages, 2000* (pp. 309–19). New York, NY: McGraw-Hill School Division.

Brown, K., Cummins, J., Figueroa, E., and Sayers, D. (1998). Global learning networks: Gaining perspective on our lives with distance. In E. Lee, D. Menkart, and M. Okazawa-Rey (eds.) *Beyond heroes and holidays: A practical guide to K-12 anti-racist, multicultural education and staff development* (pp. 334–54). Washington, DC: Network of Educators on the Americas.

Chall, J. S., Jacobs, V., and Baldwin, L. (1990). *The reading crisis: Why poor children fall behind.* Cambridge, MA: Harvard University Press.

Cummins, J. (1981). Age on arrival and immigrant second language learning in Canada: A reassessment. *Applied Linguistics, 1*, 132–49.

Cummins, J. (2001). *Negotiating identities: Education for empowerment in a diverse society* (2nd ed.). Los Angeles, CA: California Association for Bilingual Education.

Cummins, J. and Sayers, D. (1995). *Brave new schools: Challenging cultural illiteracy through global learning networks.* New York, NY: St. Martin's Press.

De Klerk, G. (1998). *Virtual power: Technology, education, and community*. Long Beach, CA: Pacific Southwest Regional Technology in Education Consortium.

Delpit, L. (1998). *Other people's children: Cultural conflict in the classroom.* New York, NY: The New Press.

Dewey, J. (1963). *Experience and education.* New York, NY: Collier Books.

Education Week (May 10, 2001). Dividing lines. In *Technology Counts 2001*, pp. 12–3. Retrieved May 24, 2002 from http://www.edweek.org/sreports/tc01/tc01article.cfm?slug=35divideintro.h20&keywords=Digital%20Divide.

Fielding, L. G. and Pearson, P. D. (1994). Reading comprehension: What works. *Educational Leadership, 51*(5), 62–8.

Freire, P. (1970). *Pedagogy of the oppressed.* New York, NY: Continuum.

Freire, P. and Macedo, D. (1987). *Literacy: Reading the word and the world.* South Hadley, MA: Bergin and Garvey.

Geva, E. (2000). Issues in the assessment of reading disabilities in L2 children — beliefs and research evidence. *Dyslexia, 6*, 13–28.

Klesmer, H. (1994). Assessment and teacher perceptions of ESL student achievement. *English Quarterly, 26*(3), 8–11.

Kourtis-Kazoullis, V. and Skourtou, E. (in press). The Internet and English language learning: Opening up spaces for constructivist and transformative pedagogy through sister-class networks. In J. Cummins and C. Davison (eds.) *International Handbook of English Language Teaching (Vol. 2)*. Norwell, MA: Springer.

Krashen, S. (September/October/November, 2001). Are children ready for the mainstream after one year of structured English immersion? *TESOL Matters, 11*(1), 4.

Kwan, A. B. and Willows, D. M. (December, 1998). *Impact of early phonics instruction on children learning English as a second language.* Paper presented at the National Reading Conference, Austin, Texas.

Manzo, K. K. (May, 2001). Academic record. In *Technology Counts 2001. Education Week on the Web*. Retrieved April 15, 2002, from http://www.edweek.org/reports/tc01article.cfm?slug=35academic.h20.

McWilliam, N. (1998). *What's in a word? Vocabulary development in multilingual classrooms.* Stoke on Trent: Trentham Books.

National Assessment of Educational Progress (2001). *The nation's report card: Fourth grade reading 2000.* Washington, DC: National Educational Goals Panel. Retrieved from http://nces.ed.gov/nationsreportcard.

National Reading Panel. (2000). Teaching children to read: An evidence-based assessment of the scientific research literature on reading and its implications for reading instruction. Washington, DC: NICHD.

Newman, D., Griffin, P., and Cole, M. (1989). *The construction zone.* Cambridge: Cambridge University Press.

Snow, C. E., Burns, M. S., and Griffin, P. (1998). *Preventing reading difficulties in young children.* Washington, DC: National Academy Press.

Tharp, R. and Gallimore, R. (1988). *Rousing minds to life.* Cambridge: Cambridge University Press.

Vygotsky, L. (1962). *Thought and language.* Cambridge, MA: Massachusetts Institute of Technology Press.

Vygotsky, L. S. (1978). *Mind in society: The development of higher psychological processes.* Cambridge, MA: Harvard University Press.

Wells, G. (1999). *Dialogic inquiry.* Cambridge: Cambridge University Press.

7 Redesigning the Foreign Language Classroom: A Critical Perspective on Information Technology and Educational Change

Michael K. Legutke

INTRODUCTION

During the last decade, many teachers and students have become involved in educational endeavors that have fundamentally changed the way we have traditionally gone about teaching a foreign language. New communication technologies allow — with just a mouse click — access to target language speech communities and to vast amounts of foreign language text. The growing mobility of European citizens — mirrored, for instance, in numerous exchange programs, language travel courses, and tourist trips — has vastly increased the chance for European students to be exposed to foreign language encounters beyond the school classroom. These changes challenge our self-concept as foreign language teachers, because, much more than in the past, we are now called upon to redefine our roles as educators, since we need to mediate between the world of the classroom and the world of natural language acquisition. Consequently, we are challenged to reshape our notion of the classroom, which, however, does not mean that we have to start from scratch. On the contrary, design features of the reshaped language classroom can already be found in the daily routines and the innovative projects of teachers and students in many parts of the world. In focusing on three such projects, I would like to foreground some of the features of the new classroom as it is taking shape and comment on some of the challenges such emerging changes pose for all of us.

In each of these classrooms, IT is being used in a particular way and in a particular context. In basing my perspective of the potential of IT on these classroom worlds, I hope to provide a platform for a constructive dialogue about how to best meet the challenges ahead of us. In the first project carried out with young learners, the computer is used as a writing tool to construct texts for a presentation and a tool to obtain simple information, whereas in the second case it functions as a research tool, allowing learners to leave the limited world of a textbook to explore more complex content. In the third case, the previous functions are utilized in addition to the computer's potential as a tool for intercultural communication. Seen together, the three teachers offer interpretations of their own practice and thus explicitly and implicitly present their theory of a new classroom. For this reason, these classrooms should not be seen as pedagogical models to be replicated in different contexts. They should rather be seen as a starting point for a debate on what is locally feasible for both teachers and learners, as arguments that might challenge us to reinterpret our own practice.

What do we mean by *classroom*? Before I take the reader into the three German classrooms, he or she might imagine entering such a room in his or her world by taking a moment to watch and listen in, peeking over the shoulders of the "inhabitants." During this excursion through the world of your imaginary classrooms, it is also worthwhile to take a look at the objects within this room, those facets of which learning environments are composed, such as tables, chairs, walls, floors, colors, forms, decorations, and exhibits.

The significance of the room in which one learns a language has been almost completely ignored by mainstream educational linguistics. Even if the space for learning is not explicitly thematized, every teaching methodology, every approach nevertheless makes indirect statements and certain demands regarding the participants' space for action. Michael Breen (1985) talks about a classroom's "culture." This culture is not only created through the interaction between students and their teachers but is in fact a manifestation of the physical space itself. It is determined by a certain place, a building, by walls, etc., which then in turn refer to larger, societal contexts: conventions, values, guidelines, building regulations and budget priorities in political arenas, and, of course, to concepts of teaching and learning. Nevertheless, despite its outward cultural dependencies limiting the space for action, the classroom is always simultaneously used and created by its inhabitants. Without ignoring the constraints, which we could all expound on with many examples from personal experience, it is more interesting and productive for the current debate to focus on the other side of the classroom's culture, the part that is created by its participants. Thus, the following cases are presented here in such a way that the teachers' voices can be heard and their viewpoints examined (cf. Bailey and Nunan 1996).

CLASSROOMS IN ACTION: THREE CASE STUDIES

English just around the corner: The classroom as a place of encounter

I met the teacher whose classroom the reader is now entering in March 1998. On my way back from the TESOL Convention, while waiting for my cab at the airport in Frankfurt, I was suddenly swarmed by a group of 11- and 12-year-olds. They were sixth graders who were doing an interview project at the airport, an outing that was part of their English lessons. Armed with tape recorders, camcorders, and notepads, they asked me, in English, about my travels, my job, and my professional life. After the interview they showed me their folders containing several worksheets I had developed twenty years ago. In short, I was encountering my own project (Legutke and Thomas 1999, 179–80, 202–3). The students gave me their teacher's name, and I called her in the evening. She invited me to participate in some of the follow-up lessons.

The day I went to her class, they were working on a world map, hung up on the corkboard in the classroom. The students used pins to mark where their interview partners had come from and where they were traveling. They then taught each other the proper terms for languages, countries, and nationalities: *One interviewee came from Thailand, she is a Thai, she speaks Thai, English and Chinese. She was on her way to Spain. She does not speak Spanish. She was a nice interview partner.* After that, some groups of students went to the library, which has four computers. Using them, they were to transcribe two minutes of their best interviews. Their task was to get ready to "teach" this part of the interview to their classmates. Others stayed in the classroom to produce a wall chart composed of a *Portrait of an Interesting Traveler* at the airport (among them my own portrait).

I later went back for a concluding post-airport activity, a parents' evening. For reasons of space, the class had moved to the art rooms, where the students had prepared various *action zones* such as the *Lost Property Office*, an *Information Desk,* City *Information,* etc. After a short welcome and introduction from the teacher, in German, the children presented a few interviews on video and audio-cassette, re-enacted interviews, read reports aloud, invited parents to participate in English in the action zones, and visit their exhibition where posters, glossaries, and, most importantly, books written by the students themselves were displayed. The *travel agency* zone even had access to the Internet. Students displayed their computer skills by putting together dream trips and last-minute offers for their parents. At the end of parents' night, the students, now in German, explained what they had learned during the project and what they had personally experienced in each of its phases.

During a follow-up interview, the teacher told me that she was doing the project for the fifth time in twelve years and that it represented the highlight of her work with her sixth grade class. She commented that this three-to-four-week project was textbook-based, meaning that students used the textbook when preparing the various project tasks, such as the interviews. She stressed the "realness" of the experience, which she thought to be extremely rewarding for the pupils, and its fundamentally co-operative nature. Students work in co-operative teams (three to five students) that are alone responsible for the use of media, documentation of results, and text production. When she started she was not using computers, but "now in the age of computer … text production is so much easier than in the past. We only have a few computers in the building, but many students use them at home."

When asked whether this project was unique, the teacher replied that it was certainly something she would consider to be a prospect for the future. She hopes and believes that the students will have such opportunities many more times as they work their way up to the tenth grade, in order to test the boundaries and the breadth of their skills in real-life communicative situations such as encounters with visitors in the classroom, on field trips, and of course in their own free time. The project should enable and encourage them to go out and look for such real-life situations, that is, to seize such learning opportunities without teacher assistance. Another aspect pointing to the "way of the future," in her opinion, is the recognition that finding and/or creating a target language environment is not only the teacher's job but rather that the children have an important part to play as well through their own texts and those collected, and as teachers themselves (outside the classroom).

When asked how she felt about going back to the normality of working with the textbook, she responded, "Why should that be a problem?" She believes those somewhat difficult or cumbersome units of the textbook are only then made plausible after the uplifting experience of having succeeded in a real-life situation. Moreover, she felt challenged to come up with one or more simulated real-life situations for each textbook unit; for example, in *Airports in the Classroom*, communicative tasks included doing a skit, writing a poem on one's own, and browsing the web for information about a dream trip in the summer. "No, my kids don't moan when we start up with the textbook again," she concluded.

Questions arising from the visualization of this classroom relate to its design features and its borders, including: Where does this classroom start, and where does it end? What is the specific role of the space between the four walls where pupils and teacher meet five times a week, and how does it relate to the world beyond? These questions are even more pertinent to the second exemplar, which follows.

Exploring native American culture: The classroom as a place of co-operative research

The students of the second classroom are two years older (cf. Schocker-v. Ditfurth 2001). In eighth grade, they are fourteen years old and in their fourth year of learning English at school. Most curricular guidelines in Germany require that a considerable amount of class time be devoted to the US, and all German textbooks offer a unit on Native Americans. The teacher whose classroom the reader is about to enter felt that this unit was not only terribly outdated but also lacked depth in its treatment of the topic. Furthermore, it promoted a stereotyped view of this aspect of American culture. For this reason, she decided to supplement the textbook with an Internet research project. She divided the class into six project groups of five students each. Each group received an Internet jump-site that would help them to explore one aspect of the topic in greater detail. They were given two class periods to search the Internet for information on their sub-topic and one period to prepare a presentation of their research results. All groups continued their web research at home. In school, they used the computer lab and the library as well as the classroom to work on this complex task. Special attention had to be given to the quality of the presentations, which had to be rehearsed and which would be evaluated by the teacher and the other students. The students were made aware that they were functioning as teachers during their presentations. In the role of teachers, they were responsible for the learning of their peers.

The following two excerpts are taken from a video documentation of the whole project and give some insights into the group work and the presentation of one of the groups. Five students have successfully done substantial research on the web on California Inter-Mountain Cultures. After the research phase, two students work on the role of totem poles, and three students focus on clothing and housing. Using instructions found on the Internet, the latter have decided to build a model thatched home (a *wickiup*) from willow rods and dried grass for their presentation. Throughout their work on the *wickiup*, while negotiating how to build it best, the students communicate in English:

Excerpt 1: Group work

S1: [while tying the rods together] *Hold it. Put it around.*

S2: *I help you, I want to help.*

[Working in silence for about a minute, student 1 puts some longer grass into the tepee.]

S2: *What do you do there, this is my way!*

S3: *Small grass into the thing.*

S1: *Yeah, into the thing, put in the small grass.*

S2: *Put this grass out, and put the small in, put the small in, put the big grass out and the small in.*

S3: *This this this we make the roof.*

Excerpt 2: Presentation

All five students begin the presentation by holding up a map drawn on a large sheet of butcher paper to locate the Indian tribe they have been working on. The poster also has several descriptive text passages, historical photos, and graphs taken from the net. Throughout the presentation, the pupils do not refer to any notes. They have practiced their lines beforehand and improvise their text. The sub-group who worked on the *wickiup* steps forward.

S1: [wears self-made Indian headband with an eagle pattern on it] *Now I'm going to tell you about their clothing. The Indians had mostly nothing on or they had* [he drops his pants to reveal a loin cloth made from leather; laughter]

S1: *Now we are going to the housing.*

S2: [picks up the *wickiup* and holds it up]

S1: [points to the structure] *This frame was built out of willow poles, this way it was built out of willow poles.*

S2: [passes the *wickiup* on to S1 holding it up]
For the roofs the Indians took grass. Put them all together with string. [She holds up a small bundle of grass and demonstrates the tying process]
S3: [points to the *wickiup*] *This is an [ab]* (the students want to use the word 'abode') *and the [ab] is the home from the California Indians. In the, in an [ab] there lived a family, the whole family.*
S2: *Sometimes they had a hole in the top* [points to the top of the tepee]
S1: *Up here* [points to the spot]
S2: *because of the fire and that came out of the smoke, when they cooking, were cooking.*

The teacher believes that her pupils are willing to "play the game" of speaking English even if she is not listening in during the group work, because they know that their presentations will be assessed by her and by the other pupils. For her, the unobserved time in the group functions as an important training ground where pupils might gain confidence in the use of the target language. In addition, she points out that the quality of content and task has an important role to play when it comes to motivating the pupils. Finally, she remarks that the efforts and achievement of her learners need to be understood within the framework of the complete project cycle: working with the textbook, going beyond the textbook by means of the Internet, working in thematically focused groups, doing research, preparing the presentation using different media and forms of creative expression, and evaluating outcomes. She finds such a work format very demanding but extremely rewarding at the same time.

Again, questions might be raised about the design features of this classroom, including the role of IT, in particular, questions about how to define the borders of this learning environment. I will return to these questions in greater detail after my third case, which consists of two rather complex email projects.

Virtual connections: The classroom as a place for intercultural learning

The two teachers whose classroom the reader is now invited into are picking up this project where the previous teachers left off (Legutke, Müller-Hartmann, and Ulrich 2000). The students in this twelfth grade Basic English course, co-taught by the two teachers within the framework of a more comprehensive research venture, are in their eighth year of English.

The project's aim was to probe the possibilities offered by the computer for intercultural learning involving an email-based correspondence project — living language links. It is well known that pen pals, and *keypals* for that matter, can only remain in touch over a longer period if sufficient consideration is given to the content of their correspondence. Usually, topics such as *house, family, dog, hamster,* and *favorite music group* wear thin relatively quickly, which leads to the first critical question: What does one write about? What do we share with another? The common ground for this project was a literary text, the well-known young adult novel *The War between the Classes* by Gloria Miklowitz (1996). The second important question is: With whom can one establish a pen/keypalship, and how does one go about establishing that initial contact? Although this is a very interesting question involving many challenging sub-questions, I will not touch upon it here. The teachers were very successful in building a virtual learning-triangle involving two classes in different parts of Germany and an English class at a school in Rosmere, Quebec, Canada. This triangle, represented pictorially in Figure 7.1, has led to an ongoing partnership among the teachers and is still functioning very successfully.

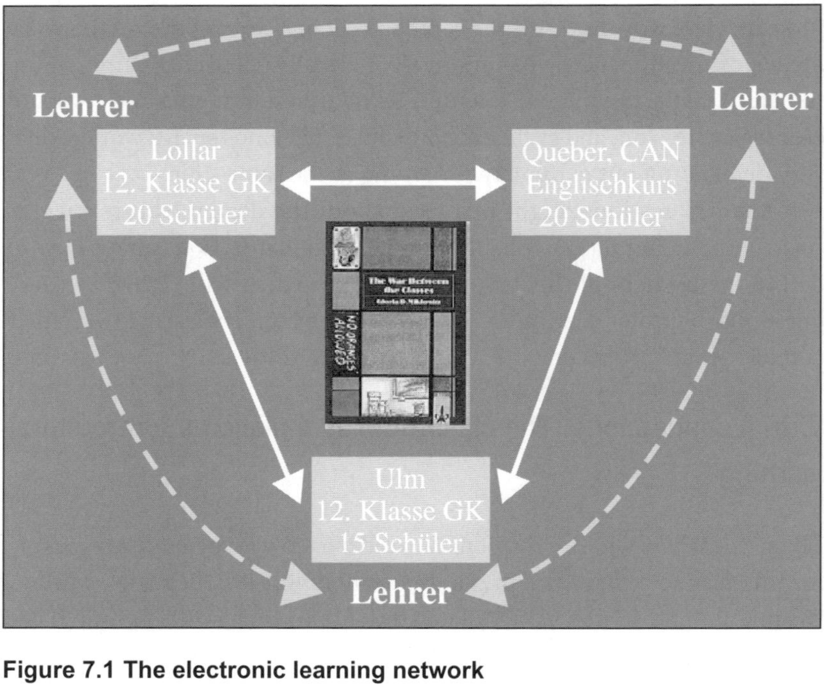

Figure 7.1 The electronic learning network

Of the many challenging issues raised by this project I, want to comment on only one: the relationship between jointly reading a young adult novel and then moving into interesting negotiations of intercultural aspects. Based on a sociology experiment, the novel deals with issues of conflict and understanding among different social classes and ethnic groups in California. After a creative introductory phase including the exchange of pictures, students were matched on a one-to-one and a group-to-group basis. Using individual reading logs that were produced by all groups, the students engaged in various discussion threads that grew out of their interests and questions. Not only did the students follow the development of the protagonists in the novel very closely, but they also related the situations in the book to their lives.

One group in Germany, for example, wrote about the social aspects of the book, pointing out that even though they "have problems with the integration of foreigners [Turks and Russians]," they stress that it is not their fault alone, but that "most of those foreigners do not have the will anymore to integrate." They closed with the following assertion:

> We think that discrimination in Germany you can find in a few places, but the main population has no problem with intercultural understanding, they live in harmony with many different ethnic groups.
> Regards Hans-Werner, Ralf, Heike, Thomas (Lollar)

Reactions from a Turkish student in the other German classroom, as well as from students in Quebec, highlight the intercultural potential of such learning networks. Bülent writes:

> Hello to Hans-Werner, Ralf, Heike and Thomas
> ... I wouldn't call that living in harmony as I am not sure what you meant by "harmony". There are serious problems between the so-called "foreigners" and German people. But this comes because both sides make mistakes. You mentioned some of them made by German people, though it is more than just settling the "foreigners" in ghettos ... On the other hand, non-German people are not open enough, especially the elderly ones. This is because they are so afraid of being assimilated, that they can't even tolerate the smallest integration because it's losing one's identity for them, whereas the problems of the young generation are mainly based on being between two cultures ... If you can say, you do not recognize many problems with ethnic minorities it is good for you in one respect, but don't you think so too, that it could be because you don't have much to do with each other — I mean here personal relationships?
> Best wishes, Bülent

A process of negotiation set in, which also involved students from Quebec. One of them compares the intra-cultural discussions in Germany to her situation:

> I found it surprising that there was a large social rift in your schools mainly between the Germans and the Turks. It is almost an exact parallel to what we have here between the French and the English. I would be interested in knowing more about some of the types of problems that arise (if any do), and also if there are many Germans with Turkish friends or vice versa. I think that although there is a definite separation in our school here between the French and the English, it can't be all that bad because I have lots of French friends.

These letters offer a number of possibilities for developing intercultural learning and understanding, but these possibilities need to be set into motion by the teacher. An outside perspective caused the participants to view their own situation — the people and the foreign elements they believed to know in their own surroundings — through a different lens and with new eyes. Thus, this outside perspective became an integral part of the foreign language learning environment for the students and teachers participating.

In the following year, the two teachers replicated the project (Müller-Hartmann 2000). This time, the basis of the project was a young adult play by Drew Hayden Taylor, *Toronto at Dreamer's Rock* (1990). Because of its topic and brevity, Taylor's drama lends itself well to this kind of project. In the play, which is only forty pages long, three 16-year-old boys, all from the Odawa/Ojibwa nation in Canada but from three different centuries, meet on a rock that is a sacred site of the nation. Taylor (1996) describes the accidental meeting of the boys, which breaks the limits of time and is accompanied by many conflicts and misunderstandings, as follows:

> I wanted to write something that explored what being 'Native' meant to different people ... three different people, from different periods in time, with different perceptions of what being 'Native' meant. ... Together the boys argue, laugh, joke, talk about girls, fight and philosophize. They question who of each of the others are and in return are questioned themselves. 500 years are put up for inspection. (p. 7)

The project lasted for three months and took a particularly exciting turn, because the three schools were joined by a survival school of Native Canadians, creating an extended electronic network, as can be seen in Figure 7.2.

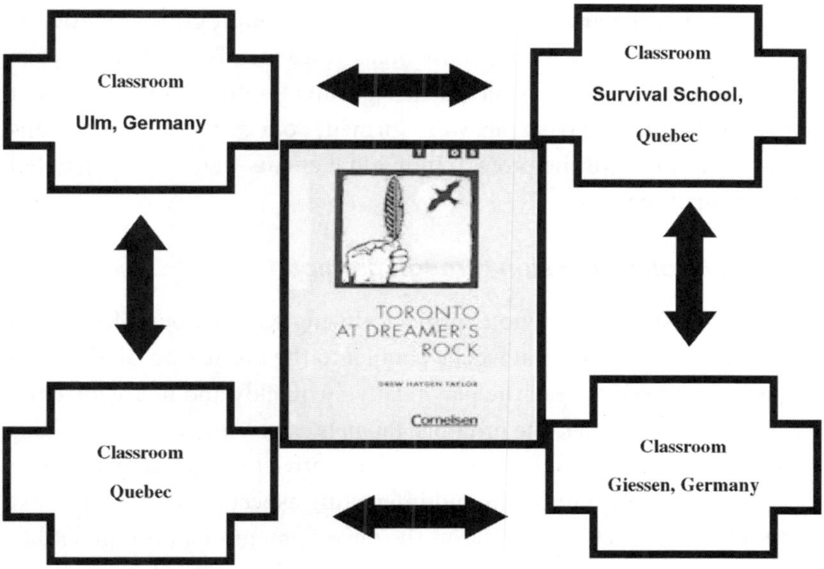

Figure 7.2 The extended electronic learning network

The three-month project was organized in three phases. In the first phase, matching cross-cultural groups were formed, and in the second, students worked with the literary text using *prompts*. These two parts together lasted about two months (twenty lessons). In the third phase, which took about one month (ten lessons), students worked on co-operative projects on Native Americans that grew out of the work on the text. The results were published on a jointly developed webpage. I would like to briefly comment on certain aspects of each phase.

Phase 1: Grouping

In order to increase the email exchange in the class and allow for a more intense form of meeting on the personal level, a group *matching* was organized instead of the more commonly found *matching* of individual students. The learners in the respective classes separated into groups according to their interests and hobbies, and chose a name for their group. A group of students interested in dancing and ballet, for instance, named themselves *Graceful Dancers*. A group from Quebec chose the name *Earth Tones*, because its members came from different cultures. Other groups chose names like *Metal Maniacs* or *Computer Freaks*. In the introduction letter to their partners, they described their common group identity as well as the special traits of the individual group members. On this basis, the groups could pick possible partner groups for co-operation, so that finally small learning triangles were formed, consisting of three groups each. In the course of the project, individual emails were written parallel to the group e-mails.

Phase 2: Book discussion through prompts

In contrast to the previous project, during which a book for young adults was read over several weeks parallel to the discussion, in this case the partners agreed to read the play relatively quickly and to organize the discussion by means of some prompts. *Prompts* can be interesting questions or provoking statements that serve as an incentive for discussion between the partners, so that important and interesting aspects of the literary text are touched on. Each class took on the development of a prompt so that *pre-*, *while-*, and *post-reading* tasks emerged that were designed and implemented not by the teacher but by learners for other learners.

Prompts:

1. *Pre-reading prompt* (Quebec):
   This is our first prompt, composed by Allison of the Earth Tones, and endorsed by the entire class as our first point of discussion:
   In fourteen hundred and ninety-two
   Columbus sailed the ocean blue.
   But raped, invaded.
   Killed and slave-traded.
   Now does he seem as good to you?

2. *While-reading prompt* (Ulm):
   What do we get to know
   a. about Rusty's world and problems in the play (first half of it?)
   b. about ourselves/about the social reality around us?

3. *Post-reading prompt* (Lollar):
   Do you think the meeting would have been different if girls (the boys' girlfriends) had come together at Dreamer's Rock?

The prompts led to an intense discussion about the multiple problems that Rusty, the protagonist, has, as well as about the social and political problems of Native Americans. The differing viewpoints around the role that Columbus played and the question of guilt with regard to the suppression of the Native Americans triggered a controversy between the partners. The incoming e-mails were answered by the respective partners or became the subject of discussion with the whole class, if the teacher thought they were especially interesting or controversial. The creative text of a Lollar group, for instance, dealing with the possible dialogue among three girls (see prompt 3) triggered a continuation of the dialogue from the Ulm group, who commented on the question of whether girls act differently from boys:

> Hello!
> Of course, we think that the atmosphere would have been different if the boys had been girls. Because girls are often able to speak more easily about personal problems. (Just think about Rusty in his inhibitions when talking about himself). We also think that Rusty's girlfriend would not use alcohol as a "tool" to solve problems.
> ... the girls would not be skeptical towards each other as boys. The topics they would talk about might also be different. Girls, often talk about fashion, make-up, music ("they are soooo cute!!!!"), films ("Titanic-mania"), horses, the future of their children and boys. Well, maybe it's not THAT bad;)
> Klaus, Mark, Werner (Computer Freaks, Ulm)

This email from the Computer Freaks in Ulm upset some of the female students in Quebec, who reacted strongly:

> We (Wendy, Susan, Ellen) find what you think girls talk about VERY insulting. You make it sound like women are narrow-minded and obsessed with superficial, materialistic bullsh*t. You are wrong. Girls that do that are immature and represent only a tiny percentage of actual women. Please do not stereotype! I NEVER talk about fashion and make-up (and I wear barely any make-up at all). When I talk about music, I talk about music talent. I couldn't care less what the members of a band look like ... And where on earth did you get the idea that girls talk about horses. It is not like the entire female population is interested in horses! I know of only two people that are.
> Well, now that I've gotten that out of my system ... let's talk about something else.
> Wendy

Wendy's email was chosen out of all the emails that arrived in Lollar for a discussion with the entire class, for which the following questions served as an orientation: What is she angry about? Can we accept her criticism? How would you answer this letter?

The results helped the students to prepare a draft for a possible answer from the whole class, and introduced questions of constructive criticism dealing with emotions, choosing the right tone, and so on. Other interesting emails were treated the same way.

Phase 3: Projects

On the basis of the aspects and questions that came from the project, groups were formed after the play discussion. They then worked on different projects and followed questions that had arisen, for instance, the presentation of Native Americans in movies or the question of what other students at their school knew or thought about Native Americans. The results were published on a website.

This complex learning environment depends on the use of computers, but the computers would have no useful function if it were not for the issues that were debated and the meaningful content with which the learners had to engage. This learning environment presents a radical expansion of the traditional classroom, and again questions arise on this classroom's prominent design features and boundaries.

DESIGN FEATURES OF THE IT CLASSROOM

The above three exemplars of learning and teaching with IT are individual accounts, faithful to a specific context and the specific participants' voices, and dependent on the personalities of the teachers and learners involved. Despite these unmistakable individual characteristics, these stories can be read as examples of good pedagogy, helping us to understand the relationship among language teaching, learning, and technology. In focusing on seven salient features of the three classrooms, I will briefly sketch my framework for such a pedagogy and thus provide my critical perspective on IT and educational change.

Situated practice

The first common design feature of the three classrooms described above is that all the learners have opportunities to interact in the target language and, through situated practice, negotiate meaningful outcomes. They are immersed in real-life communicative situations (*kommunikative Ernstfälle*) that have not been postponed to after school or relegated to the learners' private lives. These learning situations are, rather, an integral part of their classroom work: The students interact with speakers of English at the airport, they function as teachers on the basis of independent web-based research, and they communicate in a virtual learning environment. This way, learners encounter language phenomena embedded in the complexity of authentic situations rather than in a pedagogically reduced, ready-made form. One cannot plan in advance an encounter with an American pilot at the airport for a group of 11-year-olds, nor is it probable that the students from Quebec have the vocabulary lists from Unit 1 of the German textbook in their heads while writing their emails.

The challenges for the inhabitants of such classrooms looking for that real-life situation cannot be overlooked. The teachers have to be able to deal with the imponderability and complexity that this expanded space for action entails. Instead of simplifying what is to be approached and learned, as we have been trained to do in school settings, particularly in foreign language classrooms, teachers are called upon to maintain the complexity and help learners to approach and understand concepts and phenomena in the multiple complex environments in which they are found. The learners, in contrast, have to be strategically equipped so they not only pass the test in real-life situations but, in light of their interests and possibilities, autonomously add to their knowledge and skills.

The concrete forms of such real-life situations will, of necessity, vary according to a given context, just as the forms in which they can be used as learning opportunities will vary in different grade levels on the basis of the learners' age and level of knowledge. In view of today's possibilities for direct and mediated contact with foreign language (FL) speakers and encounters with non-educationalized texts, no school FL program can refuse the challenge of expanding the classroom through forms of situated practice. The spectrum of activities ranges from the incorporation of visitors in the classroom and the exploration of target-language use in the surrounding environment, to traditional and electronic correspondence with partner classes, to conducting "research" as part of international field trips, exchange visits and, of course, international school partnership programs. However, the students' private contacts and experiences have to be taken into consideration as an opportunity for classroom expansion as well.

There are three characteristics of a real-life situation that are particularly worth emphasizing:

1. It serves as a bridge between the inside of the classroom and the outside world.
2. It is of great personal importance to the participants.
3. If it challenges learners appropriately and feasibly, it can mobilize their will to achieve and create in a special way, thus maintaining a consistently high degree of motivation among participants.

Authentic audience

A second important design feature of these classrooms is that all learners get to interact with an authentic audience. Advocates of communicative language teaching (CLT) and project work have emphasized for a long time that the nature of the audience and, specifically, the nature of the speaker's relation to that audience has a crucial effect on the forms of language chosen to encode the message (cf. Johnston 1999, 55). The emergence of IT has expanded the chance for working with an authentic audience. The scope of the audience has increased considerably, as it is no longer hard for teachers to find speakers of the target language to interact with; learners can reach beyond the confines of a locally restricted environment, and they have access to communication partners of varied ages, cultures, and social backgrounds. Because of the availability of synchronous and asynchronous interaction with individuals and groups around the world, teachers no longer need to stimulate "pretend" activities but can foster meaningful communication. IT makes an audience available, yet, as you could see from the examples, we need more than IT to engage our learners with this audience.

Meaningful task-orientation

Another general feature of these three classrooms can be found in their task-orientation. They are task-driven, i.e., the sequencing of the tasks and their relation to one another depend on the main objective of each phase, a central task that we can refer to as a *target task*. Such target tasks determine the thematic points of emphasis, speech functions and vocabulary, and the selection and creation of communicative situations that need to be explored and/or trained for. The target task also determines the selection of texts. Finally, target tasks determine the necessary exercises for the teaching of linguistic, social, and strategic learning, as well as media skills. Target tasks can be complex, for example, staging a play, or they can be kept simpler with a task such as transcribing part of an interview (cf. Legutke and Thomas 1999, 166–9). The learning process is not finished with the completion of a target task, however. Many other phases follow, often initiated by the learner on his or her own, phases only made possible through the experience gained in dealing with the challenges of the target task, as the *Airport* project clearly shows.

Learner texts

The fourth design feature of these classrooms is that learners are encouraged to produce a variety of texts and engage in creative language use. The learners in the thumbnail sketches of the classrooms given here not only enter a demanding learning environment with the help of the necessary tasks but actively create this environment by producing and presenting English-language texts of their own. In the first case study, for example, not only were all the texts and written exercises from the teacher included in the students' project book but especially those texts written and collected by the students themselves. The latter were printed, copied, and illustrated in a multifariously artistic style, with glossaries, etc.

An entirely different ensemble of learner texts came from the third classroom. These texts were not written to be corrected for use of language and grammar by the teacher. More importantly, they were written by learners for learners and English-speaking partners. The texts are an expression, therefore, of various attempts made in testing the boundaries and possibilities in target-language use. In the second case, texts derived from Internet sources were collated, edited, partly rewritten, and creatively staged during the presentation. Of course, learner texts always have a diagnostic quality for teachers as well, because they bring indicators of learning progress and lack of competence to the forefront. They call attention to the learners' language deficiencies as well as to their achievements in the language. For this reason, phases of continuous evaluation represent another trait of the classrooms presented here.

Using the concept of *representational resources*, we could say that these classrooms allow for texts that derive from or stimulate interested action from socially located and culturally situated individuals — individuals who learn to use an array of semiotic resources to express themselves and thus create meaning (Cook 2000; Kress 2000; Stein 2000). This leads me to the next feature.

Resources

The fifth design feature of these classrooms is that students have access to a variety of resources and are encouraged to co-construct the learning environment. Setting one's bearings on and gaining orientation within complex learning environments through the incorporation of situated practice does not mean that pedagogically substantiated reductions and simplifications within such a classroom are ruled out. As has always been the case, their necessity is apparent when dealing with skill-oriented exercises, awareness of structures, or the introduction of genres. All three classrooms provide examples of such exercise phases. One would be mistaken to think that textbooks that are based on sequentiality and reduction lose their significance in such classrooms. However, within the scope of complex target tasks, they take on another role because they can only be partially used in a linear fashion. Thus, in the first example, the textbook proved to be the most important source for phases of language training as well as for working with collected materials and transcribing interviews. For the teacher, the textbook was utilized in the classroom in an unexpected way, because the students looked for text examples and exercise proposals in (their) first- and second-year textbooks, on their own.

The second case, however, clearly shows the limits of textbooks, because they must constantly be augmented, brought up-to-date, or even corrected. This is where computer technology and the Internet open up new possibilities and even more so if they are integrated into a school library. What you can see in the three classrooms is that learners play a decisive role not only by participating as active agents in creating a rich learning environment (by writing their own texts) but also by realizing that they can and should incorporate the texts they have gathered themselves, as shown in the first and second case studies. In this way, they seize opportunities for overt negotiation about the classroom curriculum and become responsible members of the classroom community by being accountable for the learning content.

Evaluation

The sixth feature of the design of these classrooms is that learners are induced to becoming aware of the learning process and taking responsibility for its outcome. The teacher from the first case study reports that the group got together at regular intervals to gauge their current stage or arrive at an interim balance. Such plenary meetings frequently led to phases of intense practice and training, dealing with aspects of grammar, functional language awareness, and text analysis, as well as exercises in composition and vocabulary. With regard to the teacher's own knowledge and competence, it was vital here that she remain one step ahead of the class by reviewing certain grammatical aspects of the unit as well as preparing more traditional exercises, drills, and communicative teaching strategies. Of course, creating these various communicative situations in the classroom requires planning and organization: When should the products be finished? Who can use which media and when? How should the texts be made public? Who is responsible for what?

Co-operative learning

The final design feature of these classrooms is that learners work in small, self-directed groups and act as peer teachers. The potential for the utilization of language and learning networks cannot be unlocked within the constraints of established, teacher-centered methods. A classroom defined by communicative tasks and expanded by situated practice requires co-operative, synergetic methods. One aspect surfacing in all three projects is the way in which each small group approached their objective in a different way. This is a necessary and effective form of division of labor. It is also necessary, however, that the results of each work group are made accessible to the others. This entails the taking on of teaching functions, which not only occurs in the communication between groups but also within each group, in which students were required to act as teaching aids for each other. Even if such learning environments depend on highly co-operative forms of action, they by no means rule out accustomed teaching methods. There is still room and need for frontal teaching, because the progress of the projects requires pedagogically motivated interventions, guidance, and teacher-led phases of training and knowledge transmission, without which the achievements of these groups would remain incomprehensible.

CLASSROOMS: SPACES FOR LEARNING AND LEVELS OF ACTION

In trying to label the FL classroom found in these case studies, the *workshop* metaphor might seem appropriate. Certainly, there are similarities to be found between the two: the large selection of materials, free access to different media, and the various work zones. But the classroom is more than this. It is a training ground for the present and future use of language. Training demands time, endurance, and remains inconceivable without (systematic) instruction and guidance. Furthermore, it can be seen as an observatory in which aspects of foreign language and culture are brought into focus, and since such classrooms are also places of constant design and creation, they resemble a studio as well. The term communication center comes to mind here, too. Yet despite these expansions, the classroom also remains a classroom, the place of learning and instruction as we all know it.

There is no doubt that the classrooms presented here expand learners' space for action in target language use and provide teachers with new, challenging tasks. It is also evident that the classroom's physical radius is fundamentally expanded and changed: The walls become billboards, they open up and simultaneously include the school library, the media center and/or computer lab, as well as rooms outside of the school, in the community, and even those of the target culture. The new classroom is best viewed as an ensemble of multi-functional rooms of learning, a creative arena for natural and directed foreign language learning.

The three cases highlight the fact that simple dichotomies do not work: juxtaposing learner-centeredness to teacher-centeredness, learner autonomy to teacher direction is too simple. It seems apparent that in the language classroom's expanded space for action, openness and teacher direction co-exist, just as forms of co-operative learning are compatible with phases of conventional knowledge transmission. Redesigning the language classroom is not an all-or-nothing concept, in which everything is new and the old is dismissed. Instead, the question is how the various old and newly emerging facets fit together, how the design features and their related roles and actions interplay: the training ground and the studio for text production, the communication center, and the research laboratory, and, of course the traditional classroom.

To create such a classroom, we need to take a critical perspective on our current practice. By turning the seven design features into guiding questions, we can approach our practice in a constructive way and thus become involved in the debate on educational change.

1. Does our learning environment provide for situated practice and meaningful communication? What is worth communicating about?

2. Does our learning environment allow for interaction with an authentic audience? Do not forget that other learners constitute an important element of this audience.
3. Does our learning environment involve our learners in meaningful tasks?
4. Are learners encouraged to use a variety of representational resources to create meaning and play with language?
5. Do students have access to these resources, and have they learned how to use them?
6. Are learners encouraged to turn mindfully to their own learning process and become responsible for their own learning?
7. Do learners have a chance to co-operate in co-creating the learning environment? Does the physical space allow for such co-operative behavior?

These are old questions that go far beyond the mastery of IT. Our discipline has grappled with them since the communicative turn in educational linguistics. They provide the framework for what I earlier called "good pedagogy." Language learning environments that take into account the conditions implied by these critical questions can be created by a number of different strategies and approaches, as I have demonstrated with my examples. Project formats seem to offer a powerful means to integrate them all, including IT (Legutke and Thomas 1999). The cases presented in this chapter could be read as explications of such project formats and thus of a multi-modal pedagogy that highlights the indivisibility of body and mind, of communication between the person and the world across modes, senses, and communicative practice. If we want to talk about technology, we need to discuss its use within such a framework. The most urgent problem for educators is not how to operate computers, write programs, or learn how to use the latest applications. The most crucial task is to learn how to integrate technology into the curriculum in responsible, meaningful, and effective ways. This is what redesigning the language classroom is all about. Thus, such an endeavor will be concerned, first and foremost, with issues of pedagogy; it will secondly be guided by curricular considerations. Only when the critical parameters of pedagogy and the curriculum have been established can the inclusion of IT be addressed.

REFERENCES

Bailey, K. and Nunan, D. (1996). *Voices from the language classroom. Qualitative research in second language education.* Cambridge: Cambridge University Press.

Breen, M. (1985). The social context for language learning: A neglected situation? *Studies in Second Language Acquisition, 7*, 135–58.

Cook, G. (2000). *Language play, language learning.* Oxford: Oxford University Press.

Johnston, B. (1999). Theory and research: Audience, language use, and language learning. In J. Egbert and E. Hanson-Smith (eds.) *CALL environments: Research, practice and critical issues* (pp. 55–65). Alexandria, VA: TESOL.

Kress, G. (2000). Multimodality: Challenges to thinking about language. *TESOL Quarterly, 34*, 337–40.

Legutke, M. and Thomas, H. (1999). *Process and experience in the language classroom* (4th ed.). Harlow: Longman.

Legutke, M., Müller-Hartmann, A., and Ulrich, S. (2000). Neue Kommunikationsformen im fremdsprachlichen Unterricht. In G. Fritz and A. Jucker (eds.) *Kommunikationsformen im Wandel der Zeit. Vom mittelalterlichen Heldenepos zum elektronischen Hypertext* (pp. 51–73). Tübingen: Narr.

Miklowitz, G. (1996). *The war between the classes.* Dell Publishing Company. [=1997, Stuttgart: Klett].

Müller-Hartmann, A. (2000). The role of tasks in promoting intercultural learning in electronic learning networks. *Language Learning and Technology, 4*(2), 129–47. Retrieved July 2003, from http://llt.msu.edu.

Schocker-v. Ditfurth, M. (2001). Reviving Native American culture in a German EFL classroom. *Der femdsprachliche Unterricht Englisch 35 (49)*, 23–30.

Stein, P. (2000). Rethinking resource: Multimodal pedagogies in the ESL classroom. *TESOL Quarterly, 34*, 333–6.

Taylor, D. (1990). *Toronto at Dreamer's Rock.* Saskatoon, SK, Canada [=1995, Berlin: Cornelsen].

8

Online Discussion in Secondary English: Shaping Interactive Practice for ESL Students

Kristina Love

INTRODUCTION

The interest in the use of online discussions, or electronic forums, in the area of language arts and language education has been growing steadily in tertiary contexts (e.g., Hammond 1998; Hunt 1996; Mikulecky 1998). However, the uptake of this mode of interaction has been less evident in secondary English, English teachers only gradually using online discussions to complement in-class, face-to-face discussion of a variety of issues (Bean, Cantu'Valerio, Senior, and White 1999). One of the reasons for this slower uptake may be that there are still relatively few tools available for teacher/moderators to reflect on how their online discussions may have met their purposes. Secondary English teachers, it appears, continue to explore the relatively uncharted waters of online discussion with only a few navigational aids.

In this chapter, I examine the challenges faced by one group of secondary school English teachers in Australia as they piloted their first online discussions with a group of Form 5 students (about sixteen years old). Their challenges, I believe, are representative of many that will be faced by high school teachers as they attempt to use the new technologies to enhance their students' response to literary texts. I focus in particular on the voices of a group of international students as they attempt to reconcile the academic demands of English with their social needs in this electronic mode, and on the voice of their teacher who is attempting to shape their academic and interactive practices in this new mode.

During 2000, I was invited to evaluate the piloting of online discussions as an approach to text response in the senior secondary English curriculum in a technology-rich school in Australia. Evaluation of the trial was needed for a variety of reasons, most significantly because there were no established models of online discussions being used for learning purposes in secondary English. Furthermore, the cohort of students used in the trial was a large one (200 students), preparing for a high-stakes senior school examination, and it was important to establish the value of a new approach to learning under such demanding circumstances. A final reason for the evaluation of the online discussions was that the school attracted a large number of international students, who were learning in English as a second language, and it was important to identify any benefits or disadvantages of the online mode for the learning of these students.

The evaluation identified a number of potential and actual benefits for students in using the online mode for enhancing text response in secondary English (see Love 2000; Love 2002, for a summary of these as they related to first language learners). However, such benefits depended on the teachers, who were also the discussion moderators, being aware of how to map the territory of online discussions. Discussion threads could be lengthy and difficult to track, especially for teachers who had six groups of six students to moderate, each student being required to make three postings each week for seven weeks. Under these circumstances, the potential for the online contributions to appear as chaotic was considerable.

As the external evaluator, I could provide teachers with some framework within which they could identify the existing organizational, academic, and social practices of their students as they engaged in online discussions in the pilot, and some indication of how these practices were, or could be, shaped as they moved into the second phase of their trial. To date, there had been limited investigation into the pedagogical effectiveness of interactions in computer-mediated communication (CMC). Existing studies in the area of language arts had examined online discussions with specific participants in a limited range of contexts. One such context involved teachers exchanging experiences and recommendations about adolescent literature classes as part of teacher professional development (e.g., Hammond 1998; Mikulecky 1998). Another involved college students engaged in collaborative investigation on literature (e.g., Hunt 1996). Relatively few studies however, had examined students sharing interpretations of literature at the school level (Bean, Valerio, Senior, and White 1999, being one notable exception).

The framework I was able to provide for the teachers at the trial school was based on the concepts of "purpose"' and "register" as used within Halliday's (1978) social model of language. In a Hallidayan model, the form of language is determined by the purposes its users seek to achieve in particular social contexts. Thus, the language of the Form 5 online discussions could be examined in how it served the specific purpose intended by the teachers of having students respond in a scholarly but still exploratory way to set literary texts.

Within a Hallidayan model, the register of an interaction is seen by three components: field, tenor, and mode. *Field* refers to what the language is being used to "talk" about: the topic or content of the communication. *Tenor* refers to the roles and relationships between the interactants in a communication. *Mode* refers to how the language is organized as more spoken-like or more written-like. These three features of register could account for the intellectual, the social, and the organizational techniques of online discussion facilitation, while also allowing closer examination of the language of the online contributions themselves.

In the remainder of this chapter, I illustrate how the concepts of field, tenor, and mode provide a useful map for the evaluation of online text response discussions in a secondary English context. I focus on the contributions of the international students involved in the school's trialing of such online discussions. This group of thirty international students were a distinctive group within the Form 5 cohort, receiving specialized ESL instruction. I will illustrate how the contributions of these international students can be mapped using Halliday's model of language as purposeful and his model of register. This mapping, it is hoped, will provide a means for teachers to diagnose their students' existing academic, social, and organizational skills in online discussion in English, while providing direction for the future shaping of students' skills in each of these areas.

THE ESL ONLINE DISCUSSION GROUPS

The international students were primarily from China, Japan, Indonesia, and Thailand. They were divided into six discussion groups of five or six students, their ESL teacher, Sandi, as the moderator who monitored and assessed contributions. The text being studied and discussed was a collection of short stories entitled *A Good Scent from a Strange Mountain* by Robert Olen Butler (1993). The theme of the collection was adjustment to life in a new country after the Vietnam War. This theme matched that of the novel being studied by the mainstream English students, *In the Lake of the Woods* by Tim O'Brien (1994). This anthology of short stories, rather than the novel, was selected for the ESL students because the novel was seen as a particularly challenging one, post-modernist in its structure. Sandi argued that the short stories provided more varied and engaging opportunities for discussion and clearer narrative structures for students unfamiliar with the generic structures and language features of narratives written in English.

The online discussion period around the short story anthology lasted for seven weeks, Sandi posting a new topic each week and students making an average of three contributions to the discussion of each topic. Altogether forty-two discussion threads were constructed across the six groups, most students contributing as required despite their level of language competence. Sandi could keep an electronic record of student responses to her topics that allowed her to evaluate the frequency and quality of individual student contributions as well as to track the progress of each of the groups' discussion threads. Students were explicitly advised that they would be assessed on their contributions, particularly on "how you present evidence to support your point of view, how you respect and respond to the views of others, and how actively you contribute." Students' contributions to the online discussion counted for ten percent of their mark for the term.

The online discussion thread below is typical of one posted by the ESL students during this period. All online contributions are represented as they were posted, and spelling, syntactic, and formatting inconsistencies have not been adjusted. The topic is always posted by Sandi, and student responses are listed under student names as they appear in the online discussion chain. The time and date of each contribution is also indicated.

> **<u>Topic</u> (7 Mar 9.20 am)**
> (a) Do you think Ted has adapted well to his new country? (b) Do you feel sorry for him? Explain why or why not. You can agree or disagree with other students' comments or add other questions connected with the above questions.
>
> **From — Mike M. (8 Mar 1.10 am)**
> I think that he has adapted well to his new country. Because, there is no war, he has made his new friends and he has found his new job.

From — Piter W. (9 Mar 9.30 am)
a). Yes, b). No, I don,tbecause in America ted has a comfortable lives, such asthere is no war so he doesn't have to do any fighting any more as he did before in Vietnams, he has a good job as the best chemical engineer also the peoples in his work places are good-hearted people, he also felt happy to his Son because he was born in America. My , bye...bye

From — Lydia O. (9 Mar 9.50 am)
a) I think so b) of course not I said that because Ted has a good life in America and it seem that he has no problem there, above all there is no war and his life is peaceful there. He has some friends, beside that he has a good job. He has already had a son who was born there. So I think he is happy with his new country OK that's all thank you ^_^

From — Yuni O. (9 Mar 2.30 pm)
(a). Yes, he does.

From — Cica P. (9 Mar 2.40 pm)
I think that Ted has already adapted well in his new country, but I also feel so sorry to him because his hoped to keep the Vietnamese culture cannot go on anymore. His son found the Vietnamese culture boring and never got any interested in it. That thing had made Ted very disappointed.

From — Yuni O. (9 Mar 2.47 pm)
(a). Yes, he does. He has a good live, he gets a good job, he has quite goog relationship with others (people in his new country), and the pepople also accepted him (we can know that from the name TED that given to him). (b). Not really. He has good job, he get modern live, etc. What makes me feel sorry for him is because he is living far away from the culture he used to be and yet, he is not really like the American culture. And the worst is his own son don't even show interest in vietnamnese culture (e.g. crickets fighting). Bill is more like American boy.

From — Yuni O. (9 Mar 2.20 pm)
ESPECIALLY FOR MIKE!!!!!!!!!!!! MIKE, Your answer is so short. and u didn't answer the (b) question!! do and ans it, so i can critise u

From — Mike M. (9 Mar 2.40 pm)
You said he felt happy to his son for he was born in America. Bull shit.

Ted is not happy at all with his son in as much as his son doesn't like speaking Vietnamese. Isn't it bad if you don't know to speak your mother language. I agree what you said about the environment where he works, lives. However, you haven't considered about the society in U.S.which is very bad - guns and drugs are available everywhere and there are many criminals. isn't it dangerous to live in U.S?

From — Yuni O. (9 Mar 2.50 pm)
DEAR MIKE AGAIN good improvement that your answer is "LONGER" but, can't you just use a proper English? the "Bull shit" word is just too rude, you can disagree, but not in that rude way.

From — Erik F. (9 Mar 2.20 pm)
a)Yes he does B)Maybe, because in America Ted has a good job, good live and adapted well. Beside that he never has a problem there. Ted had a son who was born in America, and also the people in his work are good people. So maybe I would like to say that Ted is pretty happy in his new country

From — Yuni O. (9 Mar 2.20 pm)
I disagree your answer. It's true that Ted has a good live, get a good job,etc. but i don't thinkk he is really happy that his son is born in America. Because that make Bill become American boy who like modern syle life. Bill also can't speak Vietnamnese. that makes Tedc feel a bit dissapointed.

From — Mike M. (9 Mar 2.30 pm)
OK!!!!! Sorry I was just trying to make a joke. Next time I will try better.

From — Cica P. (9 Mar 3.50 pm)
you guys two (yuni and mike) are crazy....don't debate about how long the answer is, talk about your fantastic opinion....oh oh oh

From — Lydia O. (10 Mar 8.20 am)
I disagree with Yuni and Cica they said that Ted miss vietnam I think that he still miss Vietnam and look after the culture of Vietnam, he had better go back to Vietnam and what for he went to America so it is useless for idiot to save him from the war. thank you I am waiting for your respond

From — Teacher (14 Mar 7.50 am)
a) I would like some reference to the story as to why your answer is

Yes, Piter. Good answer to question b)but you don't mention the concerns he has about his son.

From — Teacher (14 Mar 8.10 am)
a) Lydia, I would like you to explain why you think he has adapted well. Do you think he has any difficulty adapting to his life in USA? b) Do you think he misses Vietnamese culture and is unhappy about some of the things his son does?

From — Teacher (14 Mar 8.30 am)
Mike, you need to be careful with your language!! There will be NO abusive language. You must think of this work as part of your assessment. In answering the above question, you must base your evidence on what is in the story not what you know about America.

Clearly, there is a liveliness and engagement with the issues of the topic and with other interactants evident in this discussion thread. But what tools are available to a teacher moderator to evaluate either the individual contributions or the discussion thread as a whole? We suggest that Halliday's notion of purpose can assist in the evaluation of online discussion threads such as this; and his notion of register with its variables of mode, field, and tenor can provide a map with which teachers such as Sandi can identify what is going on in the online discussions at the organizational, intellectual, and social levels respectively.

EXAMINATION OF THE ESL STUDENTS' DISCUSSION THREADS BY PURPOSE

In broad curriculum terms, the purpose of the Form 5 English online discussions was for students to engage in exploratory, but nonetheless scholarly, reflection on aspects of a literary text with someone outside their normal friendship group. This general purpose is clearly very different from the purposes of other online discussions that may take place in discipline areas other than senior English or in contexts other than high schools. In a university text response context, for example, Hunt (1996) identifies seven genres of writing that represent seven purposes central to the conduct of his eighteenth-century university literature class, three of the most central ones being "transactionally embedded impromptu writing," "research reports," and "response to reports." In a teacher professional development context, in contrast, Mikulecky (1998) identifies five specific purposes underpinning electronic discussions between teachers of adolescent literature: (a) to *describe* teaching situations, dilemmas, and solutions; (b) to *respond* to fellow teachers' descriptions with suggestions; (c) to make *links* with the experiences of others; (d) to *support* others regarding troubling professional experiences; and (e) to *debate* controversial issues. Clearly the purposes of online discussions, even in the specific area of language arts, will vary with the participants and the nature of their task.

One key purpose identified by Mikulecky, that of debating an issue, was particularly relevant to the Form 5 online discussions, in which students were encouraged, through the wording of the teacher's prompt, to develop an argument in support of or against a particular proposition relating to the issues arising from the text. Students could either build on a shared point of view (simply adding new supporting evidence or elaborating on a point made by someone else) or contest a proposition by providing challenging evidence from the text itself.

In the online discussion thread above, we saw how Yuni, Mike, Cica, Piter, Erik, and Lydia used the online discussion to debate the issue of whether Ted had adapted well to his life in the US. Students presented evidence to support their individual points of view, as was evident in the initial contributions from Piter, Lydia, and Yuni. After a rather formulaic start, students then demonstrated considerable engagement with the views of others, bringing alternative readings of the text to contest a position put by another student, as was the case later in the discussion thread with Mike, Yuni, and Lydia's contributions. Here, the students are successfully addressing the debating purposes of the online discussion made explicit in the teacher's prompt. In the other five discussion threads in the ESL groups, students likewise recognized that they were required to debate a position about an assessment of character, drawing on specific references to the text to support their contentions.

Yet, as the teacher's final three contributions in the thread indicate, some features of the emerging discussion are less than satisfactory in the way they are achieving desired purposes. In her role as moderator, Sandi expresses concerns about the limitations of Piter's and Lydia's provision of warrants from the text to support their argumentative stance, and about the way that Mike uses language online to construct a particular relationship with others. In the post-trial interview, Sandi elaborated on these and other challenges she faced in scaffolding her international students into developing their argumentative positions online in ways that met the academic and social requirements of the subject. In the next section of this chapter, I illustrate how the registerial concepts of mode, field, and tenor can help teachers such as Sandi identify those discourse patterns in online discussions that help or hinder the achievement of specific pedagogical purposes.

Mode

Mode refers to the role that language plays in the interaction and how that language is organized to suit the channel of the communication, whether this be spoken or written or some variation on a continuum between the two (Halliday 1978). Because of its nature as electronic communication, the language of online discussion is organized such that it has some of the features of speech and some of the features of writing. In some contexts, it can permit exploratory thinking and capture ephemeral thought with all its inconsistencies and revisions. Yet, in a schooling context, it has to be organized differently from the language of face-to-face discussion, because it is more public and recordable. This aspect of mode is summed up by one of the Form 5 teachers in interview: "Online discussions are different from oral discussion — they are going to be *viewed* by others" (Mr R).

Online discussion, then, has the potential to combine the most and the least appropriate features of both the spoken and written mediums, presenting challenges for teachers who may be unsure of which features to encourage in certain contexts. The statement of purposes of the Form 5 online discussions distributed to all teachers prior to the trial suggested that these discussions would have features of both spoken and written interaction. Two itemized purposes in particular indicated that more formal written-like organization of contributor's language was required in the online discussions: "to encourage students to engage in a reflective and thinking activity with someone outside their class, normal friendship group and normal comfort zone; and to carry out a group discussion in a way that would be educationally sound but more difficult in a direct face-to-face way." One other itemized purpose, however, indicated that more spoken-like organization was required of the students' contributions to the online discussions. This item was based on the outcome for the oral component of the statewide senior English curriculum, which was that "Students should be able to express ideas and issues orally, giving considered reasons for a point of view and listening actively to the views of others." Even more specific spoken-like features of online discussions were evident in the statement of purpose distributed to teachers, which indicated that students should demonstrate the ability to do the following:

1. Follow and contribute to a discussion in a familiar setting;
2. Present evidence orally in support of a point of view;
3. Recognize and respond to cues for turn-taking in a discussion;
4. Listen actively and respond to others' views during a discussion;
5. Set own objectives for active participation in class discussion;
6. Record key points and questions raised during the discussion.

The Form 5 English teachers anticipated that their biggest challenge would lie in assisting their students to combine the more spoken-like and the more written-like requirements in their online contributions as they met the assessment goals. Interestingly, many of the international students, along with the local students, were able to accommodate features of both speech and writing in their online contributions, as demonstrated in the contributions of students like Yuni, Mike, and Lydia above. Here, some of the features of a scholarly interaction are demonstrated as students debate different perspectives on a provocative question while maintaining an interpersonal closeness and immediacy characteristic of spoken interaction. Yuni, Mike, and Lydia have been able to give "considered reasons for a point of view" and have "listened" actively to the views of others, having "recognized and responded to cues for turn-taking." Like many of the more successful international students, they have demonstrated their achievement of scholarly purposes by using syntactically well-formed sentences, formal and specialized vocabulary, and reasoning from the text as an authoritative source. At the same time, they used language for more exploratory and socially interactive purposes, incorporating more spoken-like features such as explicit and familiar terms of address (e.g., Mike, Yuni); phonemic spelling and exaggerated punctuation; abbreviations (e.g., 'u' for 'you'); questions which engage the reader as a conversationalist (e.g., "Isn't it dangerous to live in U.S?"); more explicit marking of turn-taking and the speech acts that are performed within those turns (e.g., Lydia, signing off with "OK that's all thank you"); punctuation which suggested a degree of emotional engagement (e.g., Yuni's "ESPECIALLY FOR MIKE!!!!!!!!!!!! MIKE, Your answer is so short"); and reasoning from personal as well as authorial experience.

Perhaps because the international students were used to communicating electronically on a regular basis, they used certain conventions that were more speech-like more frequently than did local students. In my interviews with international students, it emerged that most were well-off and well traveled, having been sent by their families from various Asian countries to complete their senior secondary schooling in Australia, and had been using electronic means of communicating with family and friends in the country of origin for some time. Despite the online discussions being a written record of their interactions, most of these students were undaunted by the medium and used it in a similar interpersonal mode to that for which they used email with family. This may have accounted for certain of the more spoken-like features that more regularly occurred in the postings of the international students compared with the contributions of the local students.

Yet, despite this surface casualness in the organization of some aspects of their online language, it also became evident in post-trial interviews that most international students were highly motivated to improve their English and saw the online discussion as a valuable means of rehearsing formal scholarly exchanges while engaging in more personalized communication. Given this degree of motivation, the teacher, Sandi, offered explicit guidance to ESL students who were struggling to engage with the scholarly issues of the text in an appropriate manner. Language was used by some international students in a more spoken-like mode, to explore responses that were still in the process of being shaped rather than to express already well-developed positions. The language of such threads was not very articulate, and more importantly, the logic or the forms of reasoning were not always comprehensible to other contributors. Sandi, as the teacher moderator, was quite explicit about advising students of the accessibility to a reader of their reasoning or the expression of their argument.

> **From — Teacher (28 Mar 12.40 pm)**
> Just a note to remind you of a couple of matters—Widya, you need to refer to parts from the story to show what you mean. Gobie, when you say 'interesting' it is not clear what you mean. Could you explain further. San, you refer to 'their' childhood, but we only really know about Mr.Cohen's childhood.

As with face-to-face discussion, advice contained in postings such as the one above is accessible to the whole group as an overhearing audience and not just to the individual to whom it is more directly addressed. Helping students make their reasoning and the logic of their expression accessible to others in an online environment emerged in the post-trial interviews as a major issue for all teachers, not just Sandi as the ESL teacher. All seven teachers expressed concern about how they could use the written form of the spoken word as a tool for a scholarly exploration of the class novel with a wide range of mixed-ability students. All teachers argued that they had an important role to play in regulating students to contribute in a reflective mode in all their postings. The comments of one teacher, below, are representative of this position.

> I don't want them to feel constrained by a particular structure, but I do want them to put a bit of thought into how to write it before they press the send button. I had the feeling that a lot of earlier contributions had been written off the top of the head and hadn't even been read over. And of course it's a new experience for them to be writing in a mode that doesn't have a structure built into it. (Ms D)

After one trial with online discussion, these teachers are already developing a clearer understanding of what features of the spoken mode and what features of the written mode they expect to see in their students' postings. While students are "writing in a mode that doesn't have a structure built into it," teachers can exploit the learning potential of online discussion as a more spoken-like mode that allows the kind of exploratory thinking they value at certain stages of the text-response activity. This aspect of the online mode allowed ESL students in particular to "get that fluency happening without concern for grammar and have a safe environment in which to express themselves" (Sandi, in interview). At the same time, more fundamentally written-like organization, even in this online mode, is also required of students who are preparing for a high-stakes senior secondary school examination. The Form 5 English teachers who took part in this trial are now a little clearer about the learning potentials of the spoken- and written-like modes of online discussions. In particular, they are able to identify those stages of text study units when students are required to use online discussions as a means of presenting reflective, formulated opinions in a formal mode and those stages when they can use such online discussions in a more exploratory, conversational mode, while remaining comprehensible to their interlocutors.

Following this pilot, the Form 5 teachers are now designing an approach in which a more "plastic" mode of online writing (Hammond 1998, 338) will be accepted in the earlier exploratory phases of the text study discussion, and a more formal mode will be required in the later analytical stages. This issue of the choice of mode appropriate to the stages in the exploration of a literary text leads us to a consideration of how the content of a unit of literary response is structured to unfold over a learning/teaching cycle and brings us to a consideration of field.

Field

In Halliday's model of register, *field* is that aspect that relates to the content of the interaction. In the case of the Form 5 online text response discussions, the field or academic content was determined largely by the teacher prompts that required exploration of the ethical and aesthetic issues arising out of the class texts. In the case of the ESL students, this related to the short story anthology *A Good Scent from a Strange Mountain*. Since the field of any online discussion is necessarily variable, an extended discussion of this aspect of register is not so central in a chapter on innovation in information technology. However, if mode choice varies with the stages of text study, as suggested above, so, too, would aspects of the intellectual field. A brief examination of how aspects of field are constructed in different pedagogical stages could illuminate the other registerial features.

Three key pedagogic stages had occurred before the online discussions commenced: all students had read the respective texts; teachers had used "literal" questions and "text implicit" questions (Barton 1995, 349) in the face-to-face classroom context; and through this, some building of basic "shared reading positions" (Love 2002) had been established in class groups. By the time the online discussions were initiated, teacher moderators thus presumed that students could confidently make the shift from basic comprehension questions to the more "academic knowledge" or "life knowledge" questions (Barton) used in the teacher prompts. These questions required more conceptual and analytical thinking of students, on issues that were both internal and external to the novel. In mapping the academic demands of these prompts by field, I drew on notions of text response as being either "text based" or "student based" (Barton) and on response being affectively, ethically, or aesthetically oriented (see Love 2002 for fuller details of the grid resulting from the analysis of field in these online discussions). The resulting map proved useful for Sandi in identifying which aspects of the text response field were not elicited through her prompts. This map could help her monitor her students' contributions and the design of her own prompt questions so that the full range of relevant fields is generated in future online discussions.

Equally importantly, slightly different modes were favored in the exploration of different aspects of the field. For example, when making online contributions related to their own life knowledge, students tended to use a more spoken-like mode than they did in responding to prompts concerned with text-based structures in the short story collection. Such fluctuations in the field under discussion co-occurred with and influenced the mode of response. Likewise, the registerial features of field and mode fluctuated with the tenor of the online discussion at different stages of its unfolding, as will be discussed more fully now.

Tenor

The term *tenor* refers to the role of relationships between the interactants in a communication (Halliday 1978). In online discussions, the field is jointly constructed in a public space, thus involving social interaction with other contributors. Moderators play a key role in establishing the tenor of the interaction and in maintaining productive social interactions in online discussion (Paulsen 1996). Interactions across the thirty groups of this pilot study were clearly shaped by the tone and content of the moderator's introduction (the moderator being the teacher) and by the way in which the moderator maintained expectations and regulated the social interactions throughout the discussion.

All teachers made the requirements for participation in the online discussions explicit in their introductions. Here, mindful of the fact that the online groups were not class-based groups, teachers typically welcomed students to the new group, introduced themselves, provided group members with information about the organization of the discussion, provided clear information about their expectations of students' contributing behaviors, and clarified the assessment criteria used to evaluate student contributions. These are all functions identified in the research literature as important for maintaining engagement, both in face-to-face classroom discussions (Barton 1995) and in online discussions (Paulsen 1996; Rohfeld and Hiemstra 1996). Ms P's introduction below demonstrates each of the functions outlined immediately above.

> Welcome to the Form 5 On Line Discussion for our text In the Lake of the Woods. I am the moderator of your chat group and I am looking forward to some interesting interaction with you. It is a novel that can lend itself to lots of comment and hypothesis and opinion. Please do

> not give your password to any other student – the discussion will be only between me and the 5 students in the group. The Form 5 English Teachers are able to look at the discussion in all the groups but students can only see the discussion in their allocated group. A new topic will be posted each week and we will ask each of you to make 3 or 4 contributions to the discussion of each topic. A record of what each of you says will be kept and we will be looking for the frequency and quality of your responses. We will specifically be looking for the ways in which you follow and contribute to a discussion, how you present evidence to support your point of view, how you respect and respond to the views of others and how actively you contribute. This discussion will count for 10% of the mark for the LAT for this term. Your teachers will give you occasionally some time in class to respond on line but you can of course make your responses at any time and from home if you are on line there. This is a new way of learning. We hope it is a worthwhile one for you.

All teachers provided motivating introductions of this sort that worked to engage the students, while making the teacher's role as assessor and students' roles as discussants clear right from the beginning of the interaction.

Once the discussion groups were under way, the teachers regulated the tenor of the discussion in a range of explicit ways. Because this was students' first experience with using online discussion for scholarly communication, some interpersonal experimentation became evident. Some forms of social interaction were potentially disruptive, and various teachers' methods for regulating such disruption were noteworthy. In the following interaction thread from one of the mainstream discussions, Dean, who had been a late contributor to a discussion chain that had lasted for more than a week, is attempting to make up his quota of contributions but is going about it in a potentially disruptive way. First, a student, Leoni, castigates Dean for his series of unfounded insults. Despite the concern of teachers to provide regular feedback, it is not until a week later that the teacher, Mr R, is able to check the site and to make clear to Dean the acceptable forms of interaction in this discussion.

> **From — Jess W (29 Feb 11.12 pm)**
> Covering up everything just slowed his downfall. If he didn't cover everything up then he would have gone to jail and that is more of a downfall. What he did in Vietnam was always going to bring a downfall, all he did was make it come later

From — Dean P (8 Mar 12.10 pm)
Jesse, you have no clue what you are talking about. Are you sure you have read the book mate. Sorry for any inconvenience Roger.

From — Dean P (8 Mar 1.27 pm)
G'day Robbo, I'm going to have tom disagree with what you have said. He does not gain no magic. Were did you get that form. Have you really read the book yourself. Shame,Shame,Shame.

From — Leoni W (9 Mar 2.14 pm)
Dean -before you make these accusations, about people not reading the book, and disagreeing with their opinions, we'd like to know your evidence that you have to support these theories. Otherwise, it is extrememly unfair, because I believe that everybody has put forward interesting and powerful ideas about John Wade.

From — Jess W (14 Mar 12.20 pm)
To deanNow you listen here roger. I think you need to shape up or ship out.

From — Mr R (14 Mar 4.05 pm)
Leoni, excellent point. Let's see if Dean can support what he says with references to the text.

From — Mr R (14 Mar 4.07 pm)
Dean and Jess. This is not the forum for private conversation or for non text related discussion. Please keep your comments on task and treat the others in this group, myself included, with the respect they deserve.

From — Dean P (16 Mar 1.19 pm)
I thiunk that Kathy leaving lead to his downfall. Especially when he finds out that she was murdered.

From — Leoni W (16 Mar 2.40 pm)
I disagree with Dean. Firstly, John never finds out that Kathy was murdered - in fact he never finds out what happened to her, at least not during the book. This is why the book is such a unique mystery. Secondly, John's downfall was before Kathy left. I believe his downfall was when his political campaign fell apart, or perhaps it all started was when he was at war, or he could even have been falling his whole life.

From — Dean P (24 Mar 10.01 am)
Since John's father had died and Kathy had just dissappeared, that was a real downfall for John.

Despite Leoni's attempt to regulate Dean's offensive behavior, it is only when the teacher intervenes that Dean complies. Subsequently, Dean's contributions are brief but at least presented in the required scholarly tenor and responsive to well-founded challenges by Leoni.

A similar pattern of disruption and teacher regulation occurs in the early stages of many of the mainstream groups, suggesting that online discussion involving school students is vulnerable to the same negative interpersonal dynamics as in classroom discussions. However, such disruption occurred much less frequently in the ESL groups and, typically, any disruptions that did occur were regulated by students as much as by the teacher. Recall how Yuni, in the thread we examined earlier, played a successful "surveillance" role as she reprimanded Mike for his inappropriate use of language in the teacher's absence.

From — Mike M (9 Mar 2.15 pm)
You said he felt happy to his son for he was born in America. Bull shit. Ted is not happy at all with his son in as much as his son doesn't like speaking Vietnamese. Isn't it bad if you don't know to speak your mother language. I agree what you said about the environment where he works, lives. However, you haven't considered about the society in U.S.which is very bad - guns and drugs are available everywhere and there are many criminals. isn't it dangerous to live in U.S?

From — Yuni O (9 Mar 2.50 pm)
DEAR MIKE AGAIN good improvement that your answer is "LONGER" but, can't you just use a proper English? the "Bull shit" word is just too rude, you can disagree, but not in that rude way.

In none of the ESL groups did students attempt to deliberately disrupt the purpose of the discussion. Here, complications with tenor arose largely as a result of inappropriate terminology, as evident in Mike's contribution above.

Another aspect of tenor noteworthy of examination concerns the teachers' use of feedback to build an online community. Teachers, both ESL and mainstream, were concerned to provide regular feedback to students so that "students didn't feel their contributions were being lost in cyberspace, where no one was listening" (Ms M). Teacher moderators used positive, rather than negative feedback where possible, to indicate that each contribution was being valued and to help establish the kind of scholarly community in which exploratory thinking about the novel was also valued.

> **From — Ms.D (29 Mar 7.13 am)**
> DCM to James. A most thoughtful comment. Did you get the feeling that Tim O'Brien was also writing from his own experience and did it add to the book ?
>
> **From — Ms.D (29 Mar 7.15 am)**
> DCM to Rebecca. Obviously you found the end and style unsatisfactory. Was there any solution that you felt you were led to more than others ?

Such feedback is an essential aspect of the role of a moderator (Paulsen 1996, 87; Rohfeld and Hiemstra 1996). Most teachers did as Ms D did above in that, as well as providing the *evaluation* (Cazden 1988) element in their feedback (e.g., "A most thoughtful comment"), they added personalized questions that respected the student's position but extended it further. It emerged from my interviews with students that such teacher feedback was generally welcomed by mainstream students but was valued much more highly by ESL students. The international students I interviewed wanted even more teacher intervention in the online discussion than did the mainstream students. Specifically, they wanted the teacher moderator to correct errors of expression and to provide the "right" answer to the prompt question. However, Sandi did not provide this level of feedback, firstly because of the demands on her teaching and preparation time, and secondly because she did not see her role in the online discussions as corrector of student responses. Rather, as indicated in the discussion thread with which we started this chapter, Sandi provided individualized feedback that encouraged students to support their contentions with reference to the text. The impact of the teacher's modeling of various feedback strategies had a positive flow-on effect, students in some groups providing positive feedback to other students, often as a preface to a more scholarly engagement with the issues raised, as in the interaction below.

> **From — Mathew K (17 Mar 1.28 pm)**
> James you are sounding intelligent, that is a very good way to look at it, I dont think though that Kathy should have revieled this to John

because I think he already knew, he always followed her, and was cheaking on her, just wanting to again be decived by Kathy, it did not happen so much from Kathy cheeking on John. With this in mind it is hard to see why they stayed together, Why do you think there relationship servived so long anyone (mainly James)?

From — James H (20 Mar 3.50 pm)
Thanks for your opinion Mat. I don't think that he had the slightest clue about what Kathy thought about his politics despite his stalking and constant shadowing of her life. She was trying to tell him about her feelings through her actions but he was not picking up the signs. Their relationship lasted as long as Kathy felt like it should, and in the end she was the one to end it. John would never have ended their hopeless relationship because he was obsessed with her and never wanted to lose her.

In other groups, students did provide explicit evaluation of other students' contributions, though not always in a positive or supportive form. The discussion thread below has been edited down to the contributions of two students (Phira and Judith) who assume antagonistic positions.

From — Phira C. (9 Mar 9.19 am)
For my opinion, Ted has adapted not so well in the new country. So, that's whay he always misses his home town. Maybe he should died in the war is better than someone saved him.

From — Judith Y. (9 Mar 2.30 pm)
Responding to Phira's commentI disagree with Phira's opinion saying that it would be better if Ted had died in the war. I think that Ted had a wonderful opportunity to live and start a new life in America.

From — Phira C. (16 Mar 1.11 pm)
after i've read everybody 's letter, i've found that nobody agrees with me. i'll think about Ted had a good life in USA or not again, but i also want you all my friends to think about what i'm trying to tell, and what 're the details the story always pick up with.

From — Judith Y. (16 Mar 4.10 pm)
I'm sorry that I don't agree with you Phira, but i really think that your oppinion that Ted should die in the war is rediculous... I'm sorry I said that.... hehehehheheee...

From — Phira C. (28 Mar 1.27 pm)
response to miss yani only- i know that mine is a little bit too cruel(i think), but it 's not funny !! you have to think clearly again alright hah? i know that you've been doing 'euthanasia' for a while, haven't you? so you should know that some people who disappoint in lifes want to change it. that's all my opinion to emphasise it girl.

Phira and Judith are putting into practice some of the principles taught by the teacher about structuring argument, in particular, making explicit statements of argumentative stance ("I agree/disagree with …") and demonstrating understanding of the context of the other's position as a site for developing a counter-argument ("i know that you've been doing 'euthanasia' for a while, haven't you? so you should know …"). However, the tenor of the feedback they provide to each other during this argument is rather combative ("i really think that your oppinion that Ted should die in the war is rediculous...") and could be read as hypercritical, despite the modulations ("I'm sorry I said that.... hehehehheheee..."). The ESL teacher has a key role to play here in modeling, over time, less offensive forms of feedback within an argumentative stance.

Unfortunately, scholarly student-student engagement, unmediated by the teacher, did not occur as regularly as the teachers would have hoped. All teachers in the interviews expressed their disappointment that many students saw their role in the online discussion as responding largely to the teacher's question rather than to other students' contributions. Teachers indicated that their role in scaffolding the kind of student/student interaction they wanted was crucial but difficult to achieve. Ms D's comment below was typical of many in this regard:

> Not all students knew how to respond to each other. I didn't feel I did enough to encourage them to discuss with each other. I was unsure of when to intervene, when to say things like 'Student A, Student B's made a very interesting comment. Why don't you read that and respond to him? (Ms D)

Given that the Form 5 teachers and students were working with an entirely new mode of interaction, they are in fact developing the rules of this interaction from the ground up. In their planning for the next electronic forums, the teachers are seeking to build more extensive scholarly student-student interaction in this new medium through a number of strategies. One strategy is to organize for discussion groups to meet face-to-face beforehand, to enable students to negotiate what they consider to be the most productive forms of interaction (e.g., what level of formality they want, what frequency and timing of contributions suit them). This may put more responsibility on the groups, develop student "ownership" of the groups, and allow the teacher to be more of a moderator and less of an instigator. The research literature suggests that even one well-organized face-to-face meeting can reap considerable rewards in team building and ownership (Selfe and Hilligoss 1994). Another strategy for cultivating more extensive and scholarly student-student interaction is for teachers to make explicit to students what the benefits of the online discussion would be for themselves. As one teacher put this:

> I think we need to be much clearer about advising students that they could see the discussion groups as a resource. It might be that before we start next time, we sit down and say 'Well these are the parameters and one of the things we can say is that you can use this as information that you put into your essay or your written discussion. I think that as a starting point our parameters weren't entirely clear'. (Ms F)

A third strategy the teachers are considering is to more explicitly model good student-student interaction by demonstrating how students can respond critically but courteously to other students, offer constructive compliments to other students, and articulate their meta-cognitive processes in ways that give others a clear idea of the reasoning behind their positions.

CONCLUSION

The concepts of field, tenor, and mode have proved useful in contributing to a framework for reflecting on the pilot use of online discussions for text response purposes in a high-stakes senior English curriculum. They have in particular provided a systematic means for examining the social, academic, and organizational outcomes of teachers' pedagogical choices in this new online mode. For Sandi as the ESL teacher of international students, the framework that has emerged from the pilot has helped to guide her in her planning for the next stage of her work with the online discussions, centered on a new literary text.

In relation to mode, Sandi is reflecting on what type of language organization she would explicitly require in her students' contributions at different stages of the online discussion. In the earlier exploratory stages of the online discussion, more spoken-like features of discussion may be acceptable, whereas in the later analytical stages, a more written-like mode will probably be required. Sandi may even provide dedicated opportunities for more exploratory or spoken-like forms of expression, in a separate non-assessable section of the forum that provides opportunities for what Hunt (1996, 99) refers to as "transactionally embedded impromptu writing." Such additional opportunities would allow international students to use online discussions as a means of rehearsing English language use in an exploratory mode, using the more spoken-like conventions they are familiar with. Use of this more exploratory mode could complement rather than replace more formulated online discussion contributions written in more formal, standard English. Her conceptualizing of her students' organization of "electronic language" as operating strategically on a continuum from more spoken- to more written-like modes allows Sandi to more systematically and effectively redesign her online curriculum.

The notion of field was useful for Sandi in identifying the demands of her discussion prompts and the discussion fields generated by these prompts in her students' contributions. Ideally, over a unit of work on a literary text, a full range of personal, ethical, and aesthetic responses are desirable, elicited through both *text-external* (i.e., engaging students' life knowledge) and *text internal* prompts. In conceptualizing the online discussions in these terms, Sandi can ensure that the full range of responses valued in the examination system are generated sequentially through her choice of prompt questions. The concept of field is productive not only in her systematic redesigning of her prompts but also in her redesigning of her online feedback, such that it is maximally effective in scaffolding students into making contributions across this full range.

Finally, and intimately integrated with the concepts of mode and field, Sandi's conceptualization of her role as moderator in regulating the tenor of the online discussions allows her to redesign subsequent online discussions so that they are both engaging and appropriate to a scholarly community. She can model and make more explicit the features of tenor that are required at different stages of the discussion, while modeling ways in which her international students can develop more productive ownership of the content of the discussion and critical but courteous stances towards each other. She can shape the online relationships further by providing regular and explicit feedback that actively builds a relationship among these international students that is even more supportive, both academically and interpersonally.

REFERENCES

Barton, J. (1995). Conducting effective classroom discussions. *Journal of Reading, 38*(5), 346–50.

Bean, T., Cantu' Valerio, P., Senior, H., and White, F. (1999). Secondary English students' engagement in reading and writing about a multicultural novel. *The Journal of Educational Research, 93*(1), 32–44.

Butler, R. (1993). *A good scent from a strange mountain.* New York: Penguin.

Cazden, C. (1988). *Classroom discourse: The language of teaching and learning.* Portsmouth, NH: Heinemann.

Halliday, M. (1978). *Language as social semiotic.* London: Edward Arnold.

Hammond, M. (1998). Learning through on-line discussion: What are the opportunities for professional development and what are the characteristics of on-line learning. *Journal of Information Technology for Teacher Education, 7*(3), 329–46.

Hunt, R. (1996). Collaborative investigation online: Eighteenth century literature moves to the computer lab. In Z. Berge and M. Collins (eds.) *Computer mediated communication in the on-line classroom, Vol. 2* (pp. 93–110). Cresskill, NJ: Hampton Press.

Love, K. (2000, July). *Report on the evaluation of on-line discussion at a senior secondary school.* The University of Melbourne.

Love, K. (2002). Mapping on-line discussions in secondary English. *Journal of Adolescent and Adult Literacy, 45*(5), 382–409.

Mikulecky, L. (1998). Diversity, discussion and participation: Comparing web-based and campus-based adolescent literature classes. *Journal of Adolescent and Adult Literacy, 42*(2), 84–97.

O' Brien, T. (1994). *In the lake of the woods. London: Flamingo.*

Paulsen, M. F. (1996). Moderating educational computer conferences. In Z. Berge and M.Collins (eds.) *Computer mediated communication in the on-line classroom, Vol. 3* (pp. 94–104). Cresskill, NJ: Hampton Press.

Rohfeld, R. and Hiemstra, R. (1996). Moderating discussions in the electronic classroom. In Z. Berge and M. Collins (eds.) *Computer mediated communication in the on-line classroom, Vol. 3* (pp. 94–104). Cresskill, NJ: Hampton Press.

Scholes, R. (1985). *Textual power: Literary theory and the teaching of English.* New Haven, CT: Yale University Press.

Selfe, C. and Hilligoss, S. (1994). *Literacy and computers: The complications of teaching and learning with technology.* New York, NY: The Modern Language Association of America.

9

Autonomy and Information Technology in the Educational Discourse of the Information Age

Phil Benson

INTRODUCTION

Rapid technological change is profoundly influencing both the ways in which we interact with the world and the ways in which the world interacts with us. The role of information technology (IT) in education has thus become an issue that teachers (technophiles and technophobes alike) can hardly ignore. The issue confronts us not only in the form of the workstations that we find in our classrooms and schools but also in the form of educational policies that are increasingly concerned with IT and IT-related ideas of the 'information age', and the globalized 'knowledge economy'. Formerly, educational policy limited itself to broad generalities about the goals and purposes of education and to the funding of institutions and subject areas to meet the various demands of the labour market for skilled workers and, on occasion, push the labour market in certain directions. At that time, educational policy was closely linked to the concept of 'manpower' planning. More recently, with the advent of the IT-driven information age, 'manpower' planning has been reconceptualized as 'human resource' planning and educational policy has focussed more closely on the individual as a universalized and flexible educational product. One of the consequences of this is that educational policy is increasingly moving into the domain of the classroom by focusing on the integration of IT into the curriculum and on teaching methodologies designed to produce adaptable individuals who are able to respond to the demands of a globalized knowledge economy through lifelong learning.

As a result, the pressures arising from technological change that teachers face are no longer limited to the pressure to keep up with new technologies and the technological competencies of our learners. They also include the pressure to keep up with and make sense of high-level educational policies that are increasingly concerned with the ways in which we organize our classrooms and the ways in which we teach. The problem of making sense of these policies is compounded by the fact that official educational discourse is beginning to colonize the domain of teachers' professional discourse. In the context of education policy, however, the ideas and methods that form part of this professional discourse are embedded within other discourses that often transform their purposes and meanings. In particular, the idea of *autonomy*, which has been conceptualized in teachers' professional discourse as being primarily in the interests of the learner, has tended to become an economic imperative in the education policy discourse.

The focus of this chapter is, therefore, on the ways in which relationships between IT and autonomy are articulated in the discourse of current educational policy. The context for the discussion is a number of recent policy documents advocating educational reform in Hong Kong. The issue that I wish to address is not so much whether such reforms are desirable or not, which I believe they are. It is rather the nature of the discourses that underlie proposals for reform and their implications for the day-to-day practice of teaching and learning. Emphasis on autonomy in educational policy discourse is by no means confined to Hong Kong and, although space forbids a fuller discussion of the international dimensions of educational policy, it is worth noting that autonomy has become a key policy theme around the world. Because discourses of globalization themselves imply a global discourse, educational policy is becoming increasingly homogeneous, especially in the more developed countries of the world. Thus, although the specifics of the argumentation discussed in this chapter are perhaps unique to Hong Kong, the discourses underlying them may well have a more international relevance.

EDUCATIONAL POLICY AND EDUCATIONAL REFORM IN HONG KONG

All states pursue educational policies, whether they are explicitly stated or not. Even a *laissez-faire* policy is a policy of a kind. In the sense that I use the term in this chapter, however, educational policy refers more narrowly to systematic and documented interventions by the state in the educational process. In this sense, educational policy in Hong Kong has its origins in 'manpower' planning initiatives from the early 1960s onwards (Cheng 1992). An early report (Special Committee on Higher Education 1966) recognized that "often the key to economic expansion is the planned development of manpower resources" and that "the most important means by which manpower resources can be developed is formal education and in particular formal education at the higher levels" (p. 1). A number of surveys and reports followed, and their major outcome was the massive expansion of upper secondary, university, and vocational education that took place from the 1970s through the 1990s. Evidence of these origins of educational policy remains in the name of one of the major players in current educational reform, the Education and Manpower Bureau.

From the early 1970s onwards, language education also became an issue of policy that focused both on the medium of instruction and on standards of English and Chinese in schools. Intervention in language issues was partly a response to political pressures connected to decolonization and partly a response to Hong Kong's growing perception of itself as an international city that demanded a workforce with advanced language skills. A succession of reports and interventions on the medium of instruction issue and the potential effects of the shift to Chinese-medium on standards of education has ensured that English language education is always prominent in the gaze of educational policy.

By the mid-1990s, however, widespread concern had arisen about the quality of the education system as a whole. One prominent educationalist (Salili 1996) described the Hong Kong system as competitive and examination-oriented, characterized by expository teaching and unreasonably high academic expectations. It was a system, she argued, in which it was "often difficult for the average child to survive without help in the form of individual or group tuition" (p. 90). An initial attempt at policy-driven reform — the Target-Oriented Curriculum — met with mixed success, and its failures were largely attributed to the top-down character of the initiative (Adamson and Davison 2003). More recent initiatives aiming at widespread and fundamental reform of the education system have therefore proceeded through a process of public consultation leading to the publication of a number of reports that are discussed in this chapter.

The most recent of these reports, entitled *Learning to Learn: Lifelong Learning and Whole Person Development* (Curriculum Development Council

2001), builds upon two earlier consultation documents — *Learning for Life, Learning through Life: Reform Proposals for the Education System in Hong Kong* (Education Commission 2000) and *Learning to Learn: The Way Forward in Curriculum Development* (Curriculum Development Council 2000b). Although the word *autonomy* is not used, the titles of these reports suggest that it is a key theme of the reform proposals, of which "the overarching principle is to help students learn how to learn" (Curriculum Development Council 2001, i). One of the earlier consultation documents (Curriculum Development Council 2000b) spells out what this means:

> By *Learning to Learn*, we mean that students not only learn what they are expected to learn well, but that they become better at learning new things in the future. We should help students to build up their capabilities to learn independently (e.g. creative and critical thinking, mastering of information technology, communication), to become self-reflective on how they learn, and to be able to use different ways of learning. They will then have opportunities for developing diverse ways of learning in accordance with their interests, needs and abilities in order to achieve the aims of education. (p. 3)

Learning to learn, independent learning, creative and critical thinking, reflection on learning, flexibility in learning methods, and self-determination of learning goals have all been identified as components of autonomy as they have been described in the research literature (see Benson 2001; Candy 1991 for reviews). Although academic sources are not directly cited, it is clear that the reform proposals draw on ideas that are current in this literature.

Although the use of IT in schools is not emphasized in the reports on educational reform cited above, it is the subject of an earlier document, *Information Technology for Learning in a New Era* (Education and Manpower Bureau 1998). This report is concerned more with the integration of IT into teaching and learning across the curriculum than with the need to train IT professionals, and its proposals are placed firmly within the framework of educational reform. It is also worth noting that in the quoted extract from the Curriculum Development Council (2000b, 3), IT skills are listed as one element of the capacity for independent learning, alongside creative and critical thinking and communication skills. Current policies concerning the use of IT in schools are therefore closely linked with broader policies concerning the development of autonomy in learning.

The reform proposals have also rapidly penetrated the various subject areas. Concurrently with the proposals published for consultation in the Curriculum Development Council (2000b), eight supplements were published covering key learning areas, including English language (Curriculum Development Council 2000a). In these documents, the broad principles of educational reform are developed within the context of specific curricula. Indeed, these principles were already present in the syllabuses for secondary English published in 1999 (Curriculum Development Council 1999). Here, the aims of the English curriculum are detailed as:

- to offer every student the right to a second language which provides further opportunities for extending knowledge and experience of the cultures of other people, including opportunities for further studies, pleasure, and work in the English medium; and
- to enable every student living into the twenty-first century to be prepared for the changing socio-economic demands resulting from advancement in information technology, including the interpretation, use and production of materials for pleasure, study or work in the English medium. (p. 2)

The first of these aims places the goals of language learning within the context of learning to learn goals, while the second aim places both of these goals within the context of technological change. Thus, even at a relatively early stage of the reform process, language teachers were confronted with the need to make sense of links between ideas of autonomy and ideas of the importance of IT in the context of their day-to-day practice. The nature of these links is by no means self-evident, however, and in order to understand them, we need to examine how they are discursively produced within current educational policy.

IT AS AN EXTERNAL FORCE

The origins of the emphasis on issues of autonomy in the educational reform proposals clearly lie in a perception that the current educational system does not serve learners well. This is implicit in one of the key aims of the proposals that states, "a learner-focused approach should be used to make decisions in the best interests of the students" (Curriculum Development Council 2001, i). The origins of the emphasis on IT also lie in perceptions of the utility of IT skills for learners. However, in the preambles to the various documents cited earlier, arguments based on the value of autonomy and IT skills to learners tend to be replaced by arguments that begin from a particular conception of the world, in which IT occupies a privileged position as an independent force for change. It is in these arguments that the link between autonomy and IT is forged.

The Curriculum Development Council's (2000b) description of the overall aims of a new school curriculum begins, for example, with the following statement:

> The school curriculum should provide all students with essential life-long learning experiences for whole person development in the domain of ethics, intellect, physical development, social skills and aesthetics, according to individual potentials, so that all students could become active, responsible, and contributing members of society, the nation and the world. (p. 17)

In this statement, the overall aim of autonomy is justified less by the interests of the learners than it is by the interests of society" and "the nation". This is not especially remarkable, because educational policy has traditionally identified the interests of learners with those of the state. What is new in this discourse, however, is the identification of the interests of the learners with the interests of the 'world'. Later, the aims go on to mention the "challenges of the 21st century" and "a quality curriculum for the 21st century." The aims of education, as they are represented in the new educational discourse, are, therefore, closely related to a certain conception of this world, as one that is essentially "new."

An extract from another of the consultation documents (Education Commission 2000) illustrates the ways in which this 'new' world and the challenges that it presents are conceptualized:

> We are seeing substantial changes in the economic structure and the knowledge-based economy is here to stay. Hong Kong is also facing tremendous challenges posed by a globalized economy. Politically, reunification with China and democratization have changed the ways Hong Kong people think and live. Our social structure is fast evolving and there is an urgent need to alleviate the disparity of wealth. The society is adapting its culture and mindframe to these changes. The rapid development of information technology has opened up new domains in all aspects of our lives and is creating new challenges. (p. 3)

Here, the changes to which Hong Kong society is now adapting are represented as being both global and driven by the "rapid development of IT." At the economic level in particular, the notion of a 'knowledge-based' economy links local economic changes to global phenomena that are fundamentally related to developments in IT. The outcome of this conception of our changing world is a vision of IT as the principal external force that drives the current need for educational reform.

One of the elements of this vision is a particular conception of the role of the education sector in meeting the demands of the labour market. In a speech to the Hong Kong Economic Forum in June 1998, for example, the Hon. Anthony Leung, then Chairman of the Education Commission and later Financial Secretary of the Hong Kong Special Administrative Region, discussed educational reform in the context of economic recession. Mr Leung argued that Hong Kong's potential to pull out of recession to become the "Manhattan Plus of China and East Asia" depended largely upon educational reform:

> In an industrial economy, financial capital is the determining competitive factor. In a knowledge economy, it is knowledge. Hong Kong has accumulated much financial capital. Now we have to ensure that some of our financial capital gets translated into human capital. (Leung 1998)

The metaphor of a *knowledge economy*, based on assumptions about the pre-eminent role of IT-driven international information networks in new globalized markets, leads to a particular vision of the ways in which educational policy can meet the needs of the local economy. What is required of education is less the production of individuals with specific skills than the production of a more abstract "human capital" in the form of knowledge workers who are capable of training themselves to meet changing economic needs and circumstances.

A second, more important, element in the conception of the world underlying current educational policy discourses, however, is the blurring of economics and culture that allows for the direct transfer of economic arguments and metaphors into the cultural sphere. In the education reform proposals, the human capital to which the Chairman of the Education Commission (Education Commission 2000) referred when addressing an audience of business people, is described in terms that are more likely to be acceptable to educators and to the general public:

> In the tide of changes, everyone has to meet new challenges. Adaptability, creativity and abilities for communication, self-learning and cooperation are now the prerequisites for anyone to succeed, while a person's character, emotional qualities, horizons and learning are important in achieving excellence. 'Lifelong Learning and All-round Development' is our expectation of everyone in this era. (p. 29)

The need for autonomous individuals of this kind, however, is implicitly understood to be a need that arises from a "tide of changes" that is fundamentally IT-driven. And in the new English Language syllabuses (Curriculum Development Council 1999), the opportunities for the development of knowledge and experience of the world that the acquisition of a second language offers are firmly placed within the context of technological change:

> Such extension of knowledge and experience is even more important for the twenty-first century, with the advancements in information technology bringing about not only rapid socio-economic changes and demands but also turning the world into a global village. The development of personality, values and abilities in preparation for adult life must be compatible with these changes. Analytical, critical and independent thinking, problem-solving strategies, including creativity, social competence and cultural awareness are of vital importance to our students.

In this way, reforms in the culture of education that were originally located in the perceived failures of the education system to meet the needs of learners have been transformed into reforms that meet the economic needs of an IT-driven world. This has not only placed IT at the centre of the need for reform. It has also given IT a privileged role in the process of reform itself.

IT AS AN INTERNAL FORCE

The foreword to *Information Technology for Learning in a New Era* (Education and Manpower Bureau 1998) begins by quoting the Inaugural Policy Address delivered in 1997, by the Chief Executive, the Hon. Tung Chee Hwa. In this address, the Chief Executive pledged to make Hong Kong "a leader, not a follower, in the information world of tomorrow" (Foreword). Proposals for the integration of IT into teaching and learning are thus framed within the broader socio-economic conception of an IT-driven world. The foreword goes on to argue:

> An important process in pursuit of this goal is the application of IT to enhance the effectiveness of teaching and learning so that our students will be equipped with the knowledge, skills and attitudes they need to meet the challenges of the information age.

The argument here is contentious, because there is no *a priori* evidence that the application of IT will necessarily enhance the effectiveness of teaching and learning. However, the document goes on to argue the point by analogy with the role of IT in the economy:

> IT is transforming on a global scale the way that business is conducted in both the public and private sectors. It is also quickly transforming the way in which education is delivered in schools by breaking down the traditional boundaries of teaching and learning. (p. 1)

This argument, which is based on the assumption that we now live in an IT-driven world, is then developed in ways that suggest that information technology will play a privileged role in the development of the autonomous learners that an IT-driven world demands.

The report (Education and Manpower Bureau 1998) goes on to argue, "as a powerful educational tool, IT can play a catalyst role in the transformation of school education" and outlines a vision in which the "the powers of IT" are "harnessed" in four major areas:

> - to turn our schools into dynamic and innovative learning institutions where students can become more motivated, inquisitive and creative learners …
> - to link up our students with the vast network world of knowledge and information to enable them to acquire a broad knowledge base and a global outlook …
> - to develop in our students capabilities to process information effectively and efficiently …
> - to develop in our students the attitude and capability for independent life-long learning …
>
> (pp. 1–2)

Later in this chapter, I argue that the assumption that IT has the power to bring about such changes is less than wholeheartedly supported by the research evidence. For the moment, however, let us note that the assumption that IT can act as a "catalyst" for educational reform is largely based on a twist to the more general assumption of the power of IT as a force for economic change. Once the notion of this power is accepted as an external force that demands educational reform, the argument that the same power can be "harnessed" internally in order to produce the desired reforms is readily developed.

IT AND AUTONOMY: THE RESEARCH EVIDENCE

One of the major consequences of the new educational policy to date has been the introduction of sophisticated computer laboratories and self-access language learning centers into schools. Language teachers in Hong Kong are therefore already expected to make sense of notions of IT integration and autonomy in the context of their day-to-day work, not only because they are expected to participate in the implementation of policy but also because they now tend to be timetabled to conduct lessons in computer laboratories and self-access centers. Problems of IT integration and autonomy are therefore increasingly embedded within the day-to-day practice of teaching as problems of how best to use these facilities.

Language teachers may be assisted in solving these problems to some extent by the fact that autonomy, self-access language learning, and computer-assisted language learning (CALL) have been important areas of research in Hong Kong for a number of years. Several international conferences on autonomy and self-access language learning have been held in Hong Kong over the past decade, and Hong Kong-based researchers have made an internationally recognized contribution to the literature in the field (see, e.g., Benson 2001; Benson and Voller 1997; Esch 1994; Gardner and Miller 1994, 1999; Pemberton, Li, Or, and Pierson 1996). Similarly, research on CALL has a long history in Hong Kong and has been the subject of two recent international conferences (ITMELT 1999, 2001). There is, however, a degree of tension between what we know about these issues from academic research published in Hong Kong and internationally, and the ways in which they have been represented in recent educational policy. In particular, there is a certain distortion in educational policy of both the construct of autonomy in learning as it has been developed through research and its relationship to IT.

The construct of autonomy

In educational policy, autonomy is primarily represented as an outcome of the educational process. In all the documents that have been cited in this chapter, the emphasis falls upon the role of educational reform in the production of autonomous individuals. In the literature on autonomy in language learning, in contrast, the emphasis is placed on learner control of the learning process itself. One widely quoted description of autonomy refers to it as a capacity for "detachment, critical reflection, decision-making and independent action" (Little 1991, 4). These qualities are very similar to those that are described as being desirable outcomes of educational reform in Hong Kong. The crucial difference, however, is that Little also argues that these qualities "will be displayed both in the way the learner learns and in the way he or she transfers what has been learned to wider contexts" (p. 4). In other words, the construct of autonomy in learning implies learner control over the content and process of learning as a prerequisite for the development of autonomous individuals in educational contexts. Autonomous individuals are not simply the product of supportive teaching and learning methods and arrangements. They are the outcome of learners' exercise of their developing capacities for autonomy within context of education itself.

The absence of this notion of learner control over the content and process of learning is particularly striking in the guiding principles of the new curriculum that include statements such as, "a learner-focused approach should be used to make decisions *in the best interests of students*" [italics added] and "schools have the flexibility to design their school-based curricula *to satisfy the needs of their students*" [italics added] (Curriculum Development Council 2001, i–ii). The emphasis here is on teachers and schools making decisions in the interests of their students. No mention is made of the importance of students themselves making decisions about their learning, and it is evident that autonomy is to be developed within a framework of more flexible school and teacher control. Moreover, this seems inevitable when autonomy is constructed as a socio-economic need of an IT-driven society, rather than as a capacity that serves the interests of the learners themselves.

The absence of the notion of learner control is also evident in discussions of the integration of IT into teaching and learning. In *Information Technology for Learning in a New Era* (Education and Manpower Bureau 1998), for example, the Internet is seen as a force that is deconstructing the traditional didactic role of the teacher:

> The worldwide electronic networks break down the conventional concept of teachers and textbooks being the main source of knowledge and open up new horizons for our teachers and students. Achieving more connectivity is therefore in keeping with our vision for students to become more inquisitive and creative, and to acquire a broad knowledge base and a global outlook. (p. 7)

This argument corresponds to arguments developed in the literature that it is the Internet, rather than IT per se, that offers the greatest opportunities for the development of autonomy, because it provides learners with rich opportunities for learning and interaction beyond the classroom (e.g., Warschauer, Turbee, and Roberts 1996). Here, however, it is the role of the Internet as the source of valued knowledge in the information age rather than opportunities to exercise decision-making capacities that are emphasized.

Information Technology for Learning in a New Era also identifies the Internet as a potential threat, however, because of the fear that students will be free to access "inappropriate" content. It is in addressing these concerns that a new role for the teacher in the information age is identified, for "it is exactly because students will be exposed to so much more information that the role of the teacher and facilitator becomes all the more important" (Education and Manpower Bureau 1998, 10). To clarify this role, the document goes on to argue:

> For instance, although some students may be very adept with computers, they still need guidance to adopt the correct approach in searching the right materials, so that they will not go astray amidst the many distractions on the network. Teachers need to develop in students the skills to critically and intelligently select, analyse, absorb and present the abundance of information obtainable through the use of IT. (p. 10)

We should note that this argument is in part intended to address teachers' concerns that they will ultimately be displaced by IT. In particular, it is repeatedly stressed that IT will never replace good teaching. However, it is also clear that the new role of the teacher is primarily a controlling role. In an IT-driven world, it seems logical to suppose that valued knowledge will be the knowledge that is produced and disseminated by IT and that the teacher's role as a source of knowledge will be devalued. It also seems logical to assume that the increased opportunities for learner control that the new medium offers contain an element of threat and that role of the teacher as an indirect controller of learning, charged with the task of inculcating "critical" information gathering skills, will become all the more important.

Information technology and autonomy

The argument that the integration of IT into teaching and learning promotes autonomy is also a questionable one, in the light of recent research. One important study that deserves attention in this context comes from the field of distance learning in the United States. Russell (1999) summarized the results of 355 research reports, dating from 1928 to 1998, comparing learning outcomes from technology-based distance learning programs with those from programs based on more conventional modes of instruction. Although a small number of the studies reviewed showed the superiority of technology-based instruction, these were matched by a similar number that showed the opposite. The vast majority of the studies concluded that there was no statistically significant evidence for the superiority of technology-based learning over other modes of instruction. One conclusion that has been drawn from Russell's study is that there is little or no research evidence to show that technology-based instruction is harmful to learning. An equally valid conclusion might be that we have no reason, based on research, to suppose that it enhances learning either.

In the field of language learning, much of the published research describes applications and their implementation, without any real evaluation of their effectiveness. There is also little research evidence to support the idea that the use of IT has any particular effect on learning outcomes. Although we have no review of research comparable to Russell's (1999) study, it is worth noting that one prominent CALL researcher, Chapelle (1997), has referred to the dearth of empirical studies of CALL effectiveness and adds that we still "need to specify the particularly relevant questions about CALL and to identify ways that they can be investigated through empirical research" (p. 19). We therefore lack any real evidence that IT is beneficial to learning in general or to language learning in particular, and this is perhaps why so much of the argument for the integration of IT in learning is based on metaphorical rhetoric concerning the power of IT as a catalyst for change, rather than on hard facts.

Autonomy is not an easily measurable construct, and it would perhaps be unreasonable to expect to find hard empirical evidence of the effectiveness of any particular method of promoting it. Nevertheless, several studies in the field of language learning have argued that the use of IT can in some respects enhance autonomy (e.g., Healey 1999; Kenning 1996, Warschauer, Turbee, and Roberts 1996). In a larger study of the effectiveness of various methods of fostering autonomy (Benson 2001: 111), I have argued that we need to ask two key questions that would be recast in the context of IT as follows:

1. What *opportunities* does information technology offer for learner choice and control?
2. In what ways does information technology *help* learners to take advantage of these opportunities?

When the use of IT is evaluated according to these questions, two clear conclusions emerge: The answers to the first question tend to be more positive than the answers to the second question, and answers to both questions depend very much upon the nature of the IT application and the ways in which it is used.

Many of the earlier CALL applications were based on behaviorist assumptions about second language acquisition and tend to be based on drill-and-practice or on question-and-answer routines. These applications offered learners a degree of control over the timing and pace of learning but offered little or no control over content or methods. This is an important point, because many of the CD-ROM applications currently used for language learning in schools in Hong Kong do not differ from these early applications in any significant respect other than the addition of sound and images and a degree of flexibility in the order in which items may be learned. Similarly, even the best of the language learning websites that we might point our students to tend to rely on interactive technologies that are limited to behaviorist routines.

The greater opportunities for control and choice appear to lie in the use of non-language learning applications, such as word-processors, concordances, email, and the web, that allow for control over content and methods and support learners' creative input to various degrees. Kenning (1996), for example, argues of word processing that there is "a prima facie case that by encouraging users to consider their text critically and try and make improvements, word processors are intrinsically supportive of cognitive and metacognitive autonomy" (p. 128). But although non-language learning applications offer *opportunities* for control, they do far less to help learners take advantage of these opportunities. As Kenning goes on to argue, "the effective use of electronic tools and resources assumes certain prerequisites and that unless learners already have certain attitudes, skills and strategies, they are unlikely to derive much benefit" (pp. 132–3). In other words, in order to make effective use of IT, it appears that learners must already possess some degree of autonomy in advance. Moreover, whether they possess this degree of autonomy or not is likely to depend on factors that are internal to the learner and to the teaching and learning process, and external to the use of IT itself.

It may also be worth adding that the development of autonomy is in no sense dependent on IT. In a recent review of empirical research on the effectiveness of various approaches to fostering autonomy in language learning (Benson 2001), I discussed six broad categories: (a) resource-based approaches (such as self-access), (b) technology-based approaches (such as CALL), (c) learner-based approaches (such as learner training or strategy training), (d) teacher-based approaches (involving teacher education), (e) classroom-based approaches, and (f) curriculum-based approaches (that each involve a synthesis of methods of promoting learner control and decision-making). The conclusion of this review was that the methods that have been most able to demonstrate their effectiveness are those that offer both opportunities for control over learning at a number of levels, including the content of learning, and assistance to learners in developing the abilities they need to take advantage of these opportunities. The single most effective example of the successful promotion of autonomy, proven over two decades, was identified as that developed by Leni Dam (1994) and her colleagues in Denmark, in which secondary school learners of English exercise a considerable degree of control over the curriculum and classroom activities from age eleven onwards. Significantly, Dam's model makes little or no use of technology without any apparent adverse effect on the development of autonomy among the learners.

CONCLUSION

The idea of autonomy has, in a matter of a few years, leaped from the margins of educational thought into the forefront of educational policy not only in Hong Kong but also across the world. In many ways, this is a welcome development for researchers and practitioners who advocate autonomy, because it brings ideas of autonomy to the attention of a wider audience. At the same time, it is clear that we need to be circumspect in regard to the argumentation that often underlies the theme of autonomy in educational policy. In the context of Hong Kong, arguments concerning the socio-economic value of autonomous individuals in an IT-driven world are liable to lead to a distortion of the construct of autonomy and of our understanding of the ways in which it might be fostered. Specifically, the importance of the exercise of learner control over long-term and day-to-day processes of learning is likely to be underestimated, and the capacity of IT to foster autonomy independently of this control is likely to be overestimated.

When marginal ideas become more widely accepted, it is perhaps inevitable that arguments in their favor become less rigorous or, some might argue, less dogmatic. The construct of autonomy and ideas about the best ways of fostering it are not the property of any individual or group. However, there is also a wider issue involved in the argument that I have attempted to develop that is concerned less with autonomy than with shifting boundaries of discourse and competence in education policy. The emergence of autonomy as a key theme of educational policy internationally is clearly a consequence of the presence of overarching IT-related themes of globalization, the information age, and the knowledge economy. It is, in effect, an outcome of a shift in the focus of attention of educational policy from specific 'manpower' needs to the nature of the individual as a universalized adaptable educational product. The same can be said of the link between IT in education and autonomy that is found in current educational policy. The broader process that I have been describing, however, can be characterized as one in which high-level policy increasingly articulates relationships between macro-discourses of global socio-economic change and micro-discourses of classroom practice.

The facility with which educational policy is now able to move between these macro- and micro-discourses has a number of important consequences. Educational policy no longer seeks transparently to represent the interests of the state in education but rather seeks to build community between the interests of the state and those of the learners by representing them as the common interests of a community faced with the challenge of global change. Similarly, educational reform is reconstructed as a process that no longer transparently serves the interests of learners against the more pragmatic interests of the state but rather as a process that serves the interests of all in an IT-driven world. It is this apparent community of interests, based on the assumption that IT will benefit all those who accept its 'power', that legitimizes the penetration of high-level policy into the domain of teachers' professional competence.

At this time, the eventual line of development of this process is difficult to predict. The IT policy set in 1998 was a five-year policy and in early 2004 this was reviewed a second consultation document (Education and Manpower Bureau, 2004). This document is largely positive about developments during the five-year period and employs a similar rhetoric to its predecessor, stating, for example, that IT "can be a powerful tool" for change "as Educational Reform measures steam ahead" (p. 6). The document does, however, recognize the difficulties that many teachers' are experiencing with IT and teachers will no doubt welcome the emphasis on the need for provision of more teachers' professional development opportunities. At the same time, they may be concerned by the way in which their difficulties are conceptualized as "barriers" or "obstacles affecting the use of IT in teaching and learning", and not as genuine concerns about the role of IT in Hong Kong's educational life as a whole.

It is clear, however, that the emphasis on IT in current educational policy involves shifting boundaries of responsibility, competence, and authority. In the case of IT and autonomy, we are able to see how new educational policy discourses have essentially "invaded" the domain of professional practice, shifting authority away from skilled practitioners to the policy itself. For language teachers in Hong Kong, IT and autonomy have thus become policy imperatives to be implemented rather than options to be evaluated and interpreted in the context of the experience of practice. For practicing teachers, this is likely to be a disturbing development, if only because it undermines their authority and responsibility to determine what should happen in the day-to-day practice of teaching, on the basis of their professional knowledge and competence.

REFERENCES

Adamson, B. and Davison, C. (2003). Innovation in English language teaching in Hong Kong primary schools: One step forwards, two steps sideways? *Prospect*, *18*(1), 27–41.

Benson, P. (2001). *Teaching and researching autonomy in language learning*. London: Longman.

Benson, P. and Voller, P. (1997). *Autonomy and independence in language learning*. London: Longman.

Candy, P.C. (1991). *Self-direction for lifelong learning*. San Francisco: Jossey-Bass.

Chapelle, C. (1997). CALL in the year 2000: Still in search of research paradigms. *Language Learning and Technology*, *1*(1), 19–38.

Cheng, K. M. (1992). Educational policymaking in Hong Kong: The changing legitimacy. In G.A. Postiglione (ed.) *Education and society in Hong Kong: Toward one country and two systems* (pp. 97–116). Hong Kong: Hong Kong University Press.

Curriculum Development Council. (1999). *Syllabuses for secondary schools: English language, Secondary 1–5*. Hong Kong: Government Printing Department.

Curriculum Development Council. (2000a). *Learning to learn: Key learning area, English language education*. Consultation Document. Hong Kong: Government Printing Department.

Curriculum Development Council. (2000b). *Learning to learn: The way forward in curriculum development*. Consultation Document. Hong Kong: Government Printing Department.

Curriculum Development Council. (2001). *Learning to learn: Lifelong learning and whole-person development*. Hong Kong: Government Printing Department.

Dam, L. (1994). *Learner autonomy 3: From theory to classroom practice*. Dublin: Authentik.

Education and Manpower Bureau. (1998). *Information technology for learning in a new era*. Hong Kong: Government Printing Department.

Education and Manpower Bureau. (2004). *Information technology for learning: Way forward*. Hong Kong: Government Printing Department.

Education Commission. (2000). *Learning for life, learning through life: Reform proposals for the education system in Hong Kong*. Hong Kong: Government Printing Department.

Esch, E. (1994). *Self-access and the adult language learner*. London: CILT.

Gardner, D. and Miller, L. (1994). *Directions in self-access language learning*. Hong Kong: Hong Kong University Press.

Gardner, D. and Miller, L. (1999). *Establishing self-access: From theory to practice*. Cambridge: Cambridge University Press.

Healey, D. (1999). Theory and research: Autonomy and language learning. In J. Egbert and E. Hanson-Smith (eds.) *CALL environments: Research, practice and critical issues* (pp. 391–402). Alexandria, VA: TESOL.

ITMELT (1999). Papers from the Information Technology and Multimedia in English Language Teaching Conference, Hong Kong Polytechnic University, November 1999. Retrieved June 22, 2003, from http://www.elc.polyu.edu.hk/conference/papers/index.htm.

ITMELT (2001). Papers from the Information Technology and Multimedia in English Language Teaching Conference, Hong Kong Polytechnic University. June, 2001. Retrieved June 22, 2003, from http://www.elc.polyu.edu.hk/conference/papers2001/default.htm.

Kenning, M-M. (1996). IT and autonomy. In E. Broady and M-M. Kenning (eds.) *Promoting learner autonomy in university language teaching* (pp. 121–38). London: Association for French Language Studies/CILT.

Leung, A. (1998, June 29). *The future of Hong Kong — Manhattan Plus or Venice. Human capital: The determining factor.* Speech made to the Hong Kong Economic Forum on Free Market Economy — the Way Forward. Retrieved December 1, 2001, from http://www.e-c.edu.hk/merchant/ec/docs/eng/s980629e.html.

Little, D. (1991) *Learner autonomy. 1: Definitions, issues and problems.* Dublin: Authentik.

Pemberton, R., Li, E. S. L., Or., W. W. F., and Pierson, H. D. (1996). *Taking control: Autonomy in language learning.* Hong Kong: Hong Kong University Press.

Russell, T. L. (1999). *No significant difference phenomenon.* Raleigh, NC: North Carolina State University.

Salili, F. (1996). Accepting personal responsibility for learning. In D. A. Watkins and J. B. Biggs (eds.) *The Chinese learner: Cultural, psychological and contextual influences* (pp. 85–106). Hong Kong: Comparative Education Research Centre, Hong Kong/Australian Council for Educational Research, Melbourne.

Special Committee on Higher Education. (1966). *Interim Report.* Hong Kong Government.

Warschauer, M., Turbee, L., and Roberts, B. (1996). Computer learning networks and student empowerment. *System, 24*(1), 1–14.

SECTION IV

Information Technology as Professional Development

10

Innovation and Change: Information Technology and In-service Teacher Education

David Nunan and Lillian Wong

INTRODUCTION

The purpose of this chapter is to explore English teachers' uses of information technology, their perceptions of the value and role of technology in the classroom, and their beliefs about what constitutes or hinders effective incorporation of technology in classroom practice. The chapter also aims to examine the effectiveness of an innovative, in-service teacher training program in helping teachers to implement IT in teaching. To date, comparatively few studies have investigated the processes of innovation in language teaching (for exceptions, see Henrichsen 1989; White 1988; Markee 1993, 1997; Stoller 1994). In information technology, the focus has been on student attitudes rather than on teacher beliefs. The few existing studies of teachers tend to focus on pre- rather than in-service teachers. This study fills a gap by looking at in-service teacher attitudes towards IT. It aims to provide insights into what technology brings to professional development for English teachers, to present strategies for facilitating teacher change through continuing teacher development opportunities, and to suggest directions for supporting teachers in integrating IT into the curriculum.

THE IMPACT OF IT ON TEACHING AND LEARNING

In recent years, education systems have introduced initiatives to ensure students have competitive skills for the information age. In the process, the roles of teachers and learners have changed. Teachers are no longer exclusive providers of information and knowledge. Instead, they have become facilitators in helping students to process and understand knowledge, in cultivating the innovation and creativity of students through the use of technology. Mehlinger (1995) describes this change as follows:

> The use of technology ... challenges the very relationship between students and teachers, because technology enables learners to gain control of their own learning ... The new technology provides access to information that was once under the control of teachers. (p. 94)

Dwyer (1994) reports that the use of technology in instruction encourages fundamentally different forms of interactions among students and between students and teachers, engages students systematically in higher-order cognitive tasks, and prompts teachers to question old assumptions about instruction and learning.

However, the social changes brought by technology inevitably "profoundly disturb the inertia of traditional classrooms" (Dwyer 1994, 7) and challenge established beliefs, values, and practices. Teachers are required to fundamentally change existing conceptions of teaching and learning. The use of technology in education requires teachers to modify their concepts of teaching, to reprioritize values and beliefs, and to change classroom habits (Becker 1991). In short, the deployment of technology involves change, and change entails risk.

Belief systems are both subjective and objective in nature. Kindsvatter, Willen, and Ishler's (cited in Richards and Lockhart 1996) research on teachers' belief systems states that teachers' beliefs originate from various sources, including their own learning experiences as language learners; experience of what works best in the teaching environment; established teaching styles and practices within a school; personality preferences for a particular teaching pattern, arrangement, or activity; educationally based or research-based principles; and principles derived from a particular approach or method of teaching.

In describing the historical developments in language pedagogy, Stern (1992) identifies three ways in which language pedagogy has aimed to renew and improve itself: innovation through change in teaching methods, innovation through language-related sciences and research, and technological innovation (p. 6). Stern's perspective is significant for the purpose of this review, as it encompasses the innovation of language teaching methods, learning theory, and the use of technology. While changes in teaching methods reflect to a large degree social, economic, political, or educational circumstances, they are also significantly impacted on by changes in language theories and in new perspectives on language learning (Stern 1983).

The literature on the use of technology in language learning has mainly focused on discussions and debates of the pedagogical merits of technological devices (Pennington 1996; Hanson-Smith 2000). The effective use of computers in language teaching cannot rely on the medium itself but on how it is put to use. As Garrett (1991) points out, "the use of the computer does not constitute a method"; rather it is a "medium in which a variety of methods, approaches or pedagogical philosophies of teaching may be implemented" (p. 5).

The incorporation of technology in language education, like other educational innovations, takes place at various levels and is affected by different factors. As Kennedy (1988, 331) says, "it takes place in an environment which consists of a number of interrelating systems." Within these systems, the decisive factor for effective and successful change is human. As Paisey (cited in White 1988, 116) points out:

> It is people who inhabit an institution, and an organization consists of networks of relationships between people acting and reacting on each other — thus organizations contain rational as well as non-rational elements... . Most crucially, an educational organization is operated by the persons who are themselves the instruments of change. Without their willingness and participation, there will be no change.

At the heart of an innovation in education are the teachers who implement changes in the classroom. Their attitudes and beliefs directly affect their behavior in the classroom. They are critical to the implementation of innovation. If teachers are not thoroughly involved in the implementation of innovation or they resist change, any attempts are unlikely to be successful (Marcinkiewicz 1993).

There are many reasons why innovations are adopted or rejected. They could be social, cultural, historical, political, economic, administrative, institutional, or psychological (Cooper 1989; Markee 1986; Zaltman and Duncan 1977). Potential adopters decide to adopt or reject an innovation based on their perceptions of the innovation's attributes. The five attributes of an innovation proposed by Rogers (1995) are trialability, compatibility, complexity, relative advantage, and observability. According to Rogers, the innovation will be more likely to be adopted and implemented and the rate will be faster if potential adopters can try out the innovation before adoption; if the innovation is compatible with existing practices and values, beliefs and needs, and personal and professional goals; if it is not too complex to understand and not too difficult to use; if it has an advantage relative to other innovations or the status quo; and if it offers some observable benefits. Thus, in order to carry out an innovation effectively, there must be a reasonable match between the teacher's style, philosophy, and beliefs, the working context, and the proposed innovations.

Teachers' beliefs also influence their instructional practice (Fang 1996; Pajares 1992). Past studies indicate the powerful influence that teachers' personal belief systems have on what they learn, on their decision-making, and on how they teach. Teachers tend to use technology in ways that are consistent with their personal perspectives and instructional practice, and they judge the acceptability of change by its conformance to current needs and objectives (Knupfer 1993).

Although technology has been used for instructional purposes, the main obstacles mentioned by teachers often have little to do with its pedagogical purpose (Zammit 1992). Obstacles include lack of time to find and review materials, lack of professional development to prepare teachers for the integration of technology into the curriculum, and lack of access to facilities and resources. Many teachers report a general sense of powerlessness in relation to the implementation of IT. Teachers resist using the technology because they feel coerced by those in authority to implement the innovation (Scrimshaw 1995; Terrell, Dringus, and Rendukuc 1995). Marcinkiewicz (1993) found a relationship between teachers' personality traits and the use of computers. In addition, some teachers are suspicious of the claims of the effectiveness of technology in teaching and learning (Dunkel 1987). Some, particularly the more conservative teachers, feel insecure, due to a fear that computers will threaten their roles as teachers (Adams 1985; Byron 1997). Robinson (1997) notes that teachers are worried about change because they do not understand the process of change. Ridgway (1997) points out that it is difficult to reconceptualize or change the attitudes and practices of teachers, particularly those of experienced teachers. This raises the issue of what happens to teachers when IT is rapidly introduced into an education system over a very short time. The next section deals with an example of this phenomenon in Hong Kong.

THE INTRODUCTION OF IT INTO HONG KONG SCHOOLS

As in the rest of the world, information technology has become a powerful force for social, economic, and educational change in Hong Kong. Schools are facing rapid change, both in society and in the school environment, characterized by the many changes that "move in so many different directions" (Drucker 1992, 351). In his inaugural Policy Address in October 1997, the Chief Executive set forth his vision of making Hong Kong an innovation centre for the region and promised to make Hong Kong "a leader, not a follower, in the information world of tomorrow" (Education and Manpower Bureau 1998). The First Report published by the Commission — appointed to advise him on the necessary measures to fulfill this new vision — emphasizes his desire to develop Hong Kong into "an innovation-led, technology-intensive economy in the 21st century" (Education and Manpower Bureau 1998, v).

Like governments all over the world, the Hong Kong government is spending billions of dollars to move schools into the information age with an unprecedented urgency. Hong Kong is currently undergoing holistic educational reform to transform schools to bring about the integration of IT into the classroom. A series of initiatives has been launched in schools to promote and support the implementation of IT in the school curriculum.

In September 1997, the Curriculum Development Council started to investigate different approaches to integrate IT into schools, which include the development of IT-related subjects and learning programs to enhance the IT skills and knowledge of students, the promotion of the implementation of IT across the school curriculum to empower teachers with the skills in using IT and to enable students to use IT in learning, and the use of IT as a means of educational technology to enhance teaching and learning.

The Education Commission's Reform Proposal for the education system in Hong Kong, *Learning for Life: Learning through Life* (2000), points out that "the rapid development of information technology has opened up new domains in all aspects of our lives and is creating new challenges" (p. 3). Not only the technical skills but also the pedagogical approach in using the technology to help students acquire the skills, knowledge, and attitudes to meet the challenges and to take up the opportunities of tomorrow's world are crucial to the implementation of technology in education. This paradigm shift challenges teachers to take up new roles and reshape their mode of teaching. This shift in educational goals requires teachers to change their practices and beliefs.

With the launching of the five-year plan for IT in education and major educational and curriculum reforms associated with IT integration within the whole education system, Hong Kong has entered a very exciting period of rapid expansion, development, and change in education. The challenge is not simply to apply the new technology but to manage the process of innovation.

IT AND PROFESSIONAL DEVELOPMENT: AN EXEMPLAR

One of the two overall aims of the new Hong Kong English Language Education Curriculum Framework (Curriculum Development Council 1999) is "to prepare learners for the changing socio-economic demands resulting from advancement in information technology." As part of this process, the English Centre of the University of Hong Kong was commissioned to run a fifteen-hour in-service training program for secondary school English teachers by the Education and Manpower Bureau (EMB) of the Hong Kong government. The course, *Secondary English Education (SeeIT)*, was run for three years, and about 1,800 teachers were trained.[1] Most participants were very positive about the program, and some contacted the English Centre to run workshops or courses for their schools.

The particular study which this chapter reports was based on a ten-hour program, *Integrating IT in English Language Teaching*. It was developed for a school whose principal had attended the SeeIT course and found it very useful. The school asked the English Centre to provide a similar tailor-made course for their English teachers. The course aimed to enhance participants' knowledge, skills, and strategies for using IT in teaching and build their confidence and competence in using IT.

The course was organized into four individual sessions: *Session One: Integrating IT in ELT* gave an introduction to the approaches in IT-supported English teaching and learning. *Session Two: Using the Internet in teaching and learning* created an understanding of the different functions of the Internet as well as ways of using resources from the Internet for teaching

English. *Session Three: Creating and using interactive materials for teaching and learning English* guided the teachers through important technologies to create and use the dynamic aspects of IT, including interactive web activities and web-based communication. *Session Four: Designing web-based activities* tied together all the learning and gave teachers an opportunity to apply and reflect upon the knowledge they had gained in the course.

The course was conducted in the school's computer lab once a week after the half-day school in the summer, each session being two and a half hours. A website was created for the course and all the materials were put online. Three instructors, including the researcher, all of whom had been involved with the SeeIT course, took turns to teach. While one was teaching the other two were assisting the participants in learning.

THE EFFECTS OF PROFESSIONAL DEVELOPMENT ON IT TEACHING AND LEARNING

A small-scale follow-up study was designed to examine the English teachers' beliefs about factors which supported or hindered effective classroom practice with IT, as well as their perceptions of IT implementation in English language teaching.

Four research questions underpinned the study:

1. What experience do English teachers have in using IT and teaching English with IT? How do they perceive their IT skills level?
2. What beliefs do English teachers have concerning the use and implementation of IT in teaching?
3. How do their beliefs support or inhibit the use of technology in the classroom? What stimulates or inhibits the use of IT?
4. How effective is in-service teacher training in helping English teachers to integrate information technology in their teaching?

The study focused upon teachers in a single school; hence, any supportive or negative factors identified in relation to the use of IT would reflect the ways in which the particular teachers that were studied made sense of the specific demands of that environment. However, as teachers within the same school, their access to computers as well as software, technical and administrative support was similar. Thus the study of the effectiveness of the teacher training program in helping these teachers to implement IT might provide some implications for facilitating teacher professional development in similar contexts.

The subjects, all volunteers, were two male teachers and thirteen female teachers. The majority of them were very experienced teachers, eleven out of the fifteen participating teachers having more than ten years of teaching experience. However, only four of the fifteen teachers were university graduates in English with recognized teaching qualifications, such as a postgraduate diploma of education. The others were university graduates in other subjects, with or without teaching qualifications. Five of the fifteen teachers volunteered to participate in focus group discussions. Two groups were established according to the availability of individual teachers. One group consisted of three highly experienced teachers. The other group consisted of two experienced female teachers, one of whom was non-Chinese.

Questionnaires, focus group discussions, and an in-class discussion activity were used to collect data to examine teachers' use of technology, to explore their perceptions of IT in teaching, and to evaluate the effectiveness of the in-service teacher training program in helping teachers in the implementation of IT.

The questionnaire was based on the standards and concepts developed by the Teaching, Learning, and Technology Group (TLT Group) in the *Flashlight Evaluation Handbook* (Ehrmann and Zúñiga 1997). This resource offers widely tested and accepted items and procedures for creating periodic studies of views on changes in teaching and learning practices and for examining the effectiveness of technology in education. A pre-course questionnaire, completed by fifteen teachers, was designed to identify the participants' background, their experience and level of comfort with technology, their training or professional development related to IT, their practices and beliefs in using IT in teaching, and their self-perceived IT skill levels before attending the course. A end-course questionnaire was given to all the course participants at the end of the last session of the course, to assess any changes in their attitudes and beliefs about the use of IT in teaching, their perceived IT skill level, their plans in using IT, and the effectiveness of the program in helping teachers to implement IT in their teaching. As three of the end-course questionnaires were incomplete, only twelve were analyzed.

To probe more deeply into the ways in which the teachers perceived the role and value of IT, their beliefs, the stimulus or obstacles in using technology in their teaching, and their future plans for IT implementation, at the end of both the pre-course and end-course questionnaires participants were asked to participate in end-of-course focus group discussions. Two focus group discussions, one group with three, another group with two volunteers, were conducted at the end of the course. Each discussion took between forty-five and sixty minutes.

A discussion activity was also carried out in the first session for teachers to share their views on the use of IT in education. Teachers worked in small groups of three to four and were given ten minutes to brainstorm any issues they felt made the use of IT in the curriculum either positive or negative. They were encouraged to be creative and consider all the effects that IT was having or might potentially have on their teaching situation and the learning process of their students.

TEACHER BELIEFS AND IT USE: RESULTS AND DISCUSSION

This section presents the results from the pre-course and end-course questionnaires, as well as from the in-class discussion activity and the focus group discussions. The results include information on the teachers, frequency and types of IT use, as well as beliefs about IT use.

The results of the questionnaire revealed that, not surprisingly, the teachers used computers both at home and at work, the majority (eighty percent) using computers for two hours or less at home and ten hours or less at work in a typical week. However, conducting lessons in a computer lab was uncommon. Only six out of fifteen teachers had used a computer lab in a typical week, but they did not use these facilities much (three used them for two hours or less, and three used them for ten hours or less). Teachers were not ready to use IT in the classroom, they did not know what to do in the computer lab, and they were not competent to take the class to a lab full of computers for lessons. Computers were viewed as mainly for personal use, and teachers rarely made the leap to use them in the classroom, mainly because they felt that they lacked the knowledge and skills to do so.

In general, a correlation was found between the time that teachers spent using computers and their own perceived knowledge of the different skills. Figure 10.1 shows the average time teachers spent on different tasks per week.

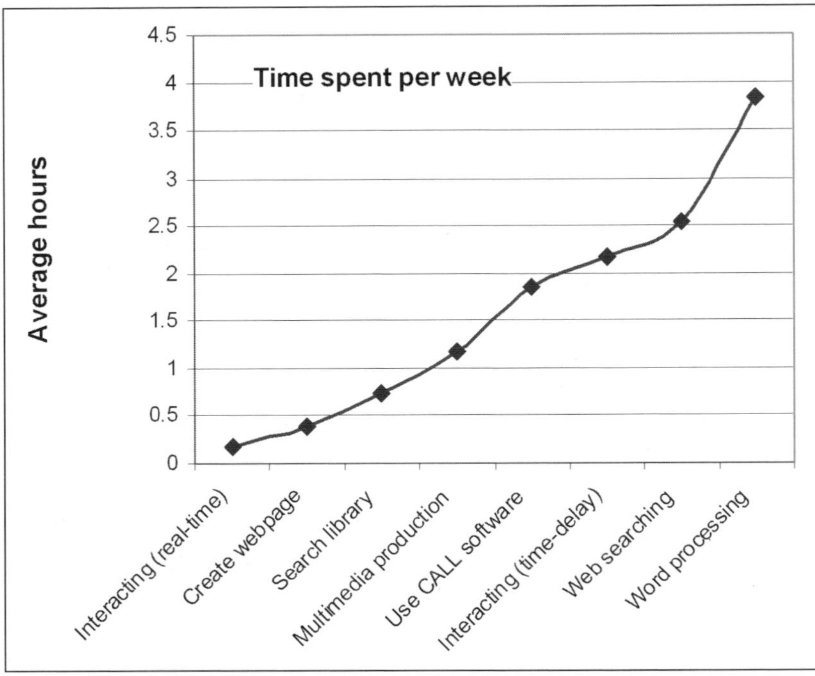

Figure 10.1 Average time spent per week on different activities using computers

As Figure 10.1 shows, on average teachers spent less than four hours per week on the most popular activity, using a word processor, and spent virtually no time on activities requiring special software or advanced skills. Teachers had a very limited view of some of the technologies listed as examples in each category of the questionnaire. For example, when asked how much time and how competent they were at using "time-delay electronic communication (such as bulletin-boards and discussion lists)," very few teachers did these things, but they felt quite knowledgeable in using email (eighty percent of teachers indicated their knowledge/ability in sending and receiving email as normal to experts users). However, email received from the teacher participants were often missing critical information such as the person's name, return email address, and subject line. Even the teachers who sent and received personal email did not know how to incorporate this technology into classroom activities.

Teachers spent relatively more time using a word processor (sixty percent spent three to over ten hours on this per week), and most felt competent in creating a word document (eighty percent indicated their knowledge/ability in creating a word processed document as normal to experts users). However, when asked to use the word processor in the course, many had problems with simple functions such as bullet points and numbering, spacing and paragraphing, tables and forms. More advanced skills such as insert comments and online marking were virtually unknown.

Other technologies such as real-time interaction were largely neglected. In contrast, in the focus group discussions, the teachers acknowledged, "students know a lot." However, after being introduced to interactive chat, teachers admitted that it was great fun participating in a chat room and they understood better why students liked it. One teacher pointed out its potential for motivating students.

When web searching was introduced in the course, most teacher participants had given little thought to the techniques used in conducting a search, and they had only used one or two search engines. While most knew how to search the web, they did not know where to find suitable CALL materials. Teachers seldom produced multimedia materials (thirty-three percent had never tried it; forty percent did it less than one hour per week). Creating or updating a home page on the Internet was also uncommon (fifty-four percent had never done it, and the other forty-six percent spent only two hours or less on doing it per week).

As can be seen from Table 10.1, teachers in general perceived that their IT skills were average, about half of them ranking themselves as three on a five-point competency scale. However, one-third also indicated anxiety.

Table 10.1 Self-perceived IT skill levels

	Anxious ——→ Very competent				
N=15	**1**	**2**	**3**	**4**	**5**
	2	3	8	2	0
	13%	20%	53%	13%	0%

Teachers with less teaching experience appeared more comfortable with computers in general. The four teachers with less teaching experience (two with under three years, and two with four to six years of teaching experience) had taken some teacher development courses to help improve their IT skills in teaching.

In relation to teachers' articulated beliefs about teaching and learning with IT, there was considerable variation in their reasons for using IT, as can be seen from Figure 10.2.

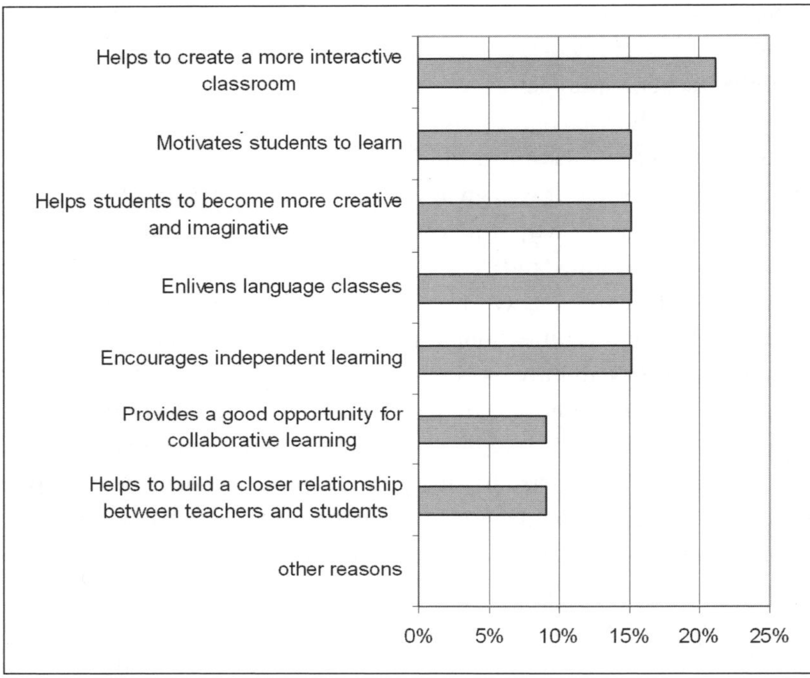

Figure 10.2 Reasons for using technology

The most popular reason for using IT was that it helped to create a more interactive classroom. Other common reasons were that the computer encouraged independent learning, enlivened language classes, helped students to become more creative and imaginative, and motivated students to learn. However, few teachers recognized the capacity of computer to connect people. Fewer than ten percent of the teachers stated that the computer helped to build a closer relationship between teachers and students or provided a good opportunity for collaborative learning.

The results show that teachers perceived the computer as an add-on tool that helped to motivate students to learn and to create a lively classroom atmosphere for learning.

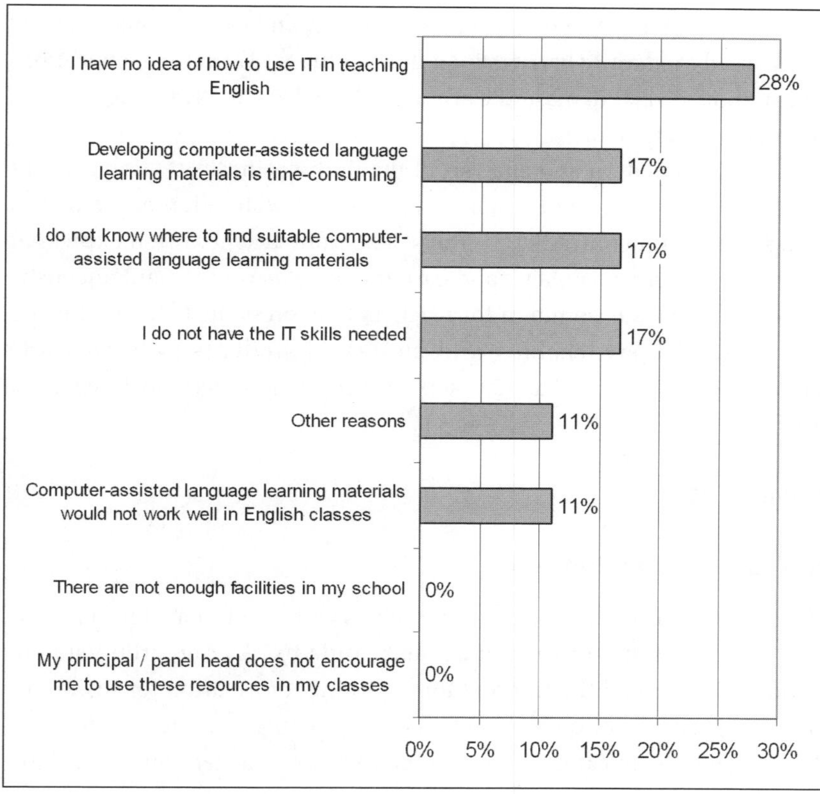

Figure 10.3 Reasons for not using technology

Figure 10.3 shows the biggest obstacle to effective IT use was lack of knowledge. Other factors hindering use of IT included time, lack of IT skills, and not knowing where to find computer-assisted language learning (CALL) materials.

The fact that no teachers indicated facilities and support from administrators was a problem that prevented them from using IT shows that the school had the infrastructure and resources required. This clearly indicates the school management's commitment provided a good environment to support teachers and students in using IT.

Further insights into teachers' beliefs about reasons for using or not using IT came from the discussion activity and focus groups. By coding key words, participant statements were assigned to ten categories. Five of these reflected advantages and five reflected barriers to IT use. Advantages were categorized as (1) resources and information, (2) communication and collaboration, (3) motivation, (4) learning, and (5) teaching. The key words used to identify barriers included (1) time; (2) training, education, and support; (3) students and learning, (4) teachers and teaching; and (5) access and equipment.

The presentation and analysis of the findings in this section attempt to capture the essence of teachers' beliefs that were often expressed in complicated and contradictory ways. The same teacher often expressed beliefs about both advantages and barriers to the use of IT simultaneously. For example, it was common for teachers to express the belief that using IT enhanced their teaching and motivated students to learn while simultaneously arguing that they lacked the time and skills to make effective use of IT.

Advantages

Resources and information

IT, especially the Internet, provided teachers with "up-to-date information" that was beneficial to them and their students. As one group stated, "students can go on the Internet for obtaining information for research." Another group mentioned, "students can be exposed to new materials." Additionally, the information on the Internet gave teachers "more teaching ideas."

Resources were perceived as not only up-to-date but also varied and comprehensive. One group stated, "students are exposed to a wider source of information and materials," and another pointed out, "different subject areas are covered" and students were exposed to materials globally. The current and varied nature of online information was connected to the need for cross-disciplinary teaching and learning.

Teachers also appreciated the ready accessibility of the information afforded by online resources. In addition to the convenience of online information, its authenticity appealed to teachers. One group pointed out that authentic materials are easily found for teaching and learning. It was felt that authentic and real-world learning increased student motivation and generally resulted in a more effective learning environment.

Communication and collaboration

Opportunities for real-world language use and meaningful communicative interaction were also seen as a significant advantage. One group of teachers asserted that the use of technology is "able to bring the outside world into the classroom." Students had the opportunities to participate in the reality of communication as it existed outside of the classroom. Another group pointed out that there was "no limit of boundary," as it involved more people and thus generates more ideas.

Motivation

The motivation offered by technology was another important aspect mentioned by teachers. Many noted that using technology is "more interesting" and "exciting" for students because it is "interactive and dynamic," students have "some different kinds of activities to do." Examples included doing research about the topic of "food" from different places, and sending e-cards on Mother's Day.

Teachers found the "visual effect" was especially motivating to students. Pusak and Otto (1997) discuss the motivating effects of multimedia, pointing out that students can be much more intellectually engaged by interacting with complex mediated programs that present language and culture in context: "They are motivated both by authentic experiences with the language and by the prospect of gaining skills that might have practical application for them" (p. 6).

Learning

Teachers indicated that technology enhanced learning: "the students' knowledge dimension will be broadened as they can browse the websites of other subjects," use of technology "promotes self-learning among students" and "interactive learning," and it "arouses creativity." It facilitated individualization, autonomy, and enhanced the learning process in general.

Teaching

In one focus group discussion, a teacher described "the computer as tutor," and in the class discussion activity, a group stated that the computer "serves as a teaching aid." Several other groups had similar views, suggesting that IT facilitated lesson preparation, as teaching materials presented in, for example, PowerPoint, could be modified easily and reused. The multimedia language center (MMLC) room also allowed students to listen to instructions and follow simple guidelines, and the presentation of grammar items was clearer to students with the help of PowerPoint.

Warschauer (1997) reminds us that technology should not be simply an add-on to the curriculum; rather it should represent an attempt at integration and place sufficient cognitive and linguistic demands on students. However, teachers in this study saw technology as a supplement to the established curriculum rather than integrating it into the curriculum. This reflects findings from Law (2000) in her study on IT implementation in Hong Kong schools. She found that the most prevalent model of IT use is that of empowering advanced education technology tools for the teacher, "resulting in more presentation oriented, didactic classroom practices … in fact not surprising but extremely worrying."

For many teachers, enthusiasm about the advantages of the use of technology is coupled with frustrations experienced in their attempts to make use of it. These experiences have led the teachers to form beliefs that raise important issues and questions about the use of technology in language learning and teaching. The next section looks at perceived barriers and challenges to using technology.

Barriers

Time

The teachers in this study saw time as the major barrier to the use of technology. While they believed that technology has a great advantage in facilitating and enhancing the curriculum, as discussed in the previous section, they also saw using it as very time-consuming. Many teachers referred to lack of available time to prepare for using technology in their teaching, and several argued that the heavy workload, especially for English teachers, made it difficult for them to find time "after a long day of teaching and marking." Thus, investing the time to reach a certain comfort level in using technology and to keep up with the changes was a big challenge for teachers, as they remarked in most in-group discussions. These beliefs suggest that the teachers in this study may be at an early stage of using technology. They may be open to investigating the potential of using technology but require release time or extra time for preparation.

Training, education, and support

Lack of IT skills was a prominent problem mentioned by the teachers. Many teachers felt that they were ill equipped to teach using technology, due to insufficient IT knowledge and training, and lack of familiarity with IT teaching. These recurring comments suggested that technical training is one area of professional development that is badly needed.

In addition to the technical aspects of using technology, many teachers pointed out the need to know about IT-based resources that they could use in their classes. Such knowledge would provide teachers with more materials for teaching. They requested information about relevant websites, activities, and materials.

Also, a number of teachers expressed anxiety and fear about knowing less than their students, some teachers commenting, "students know a lot". This echoes Hannafin and Savenye's (1993) findings that teachers resisted using computers because they were afraid of losing control over students or of "looking stupid" in front of the class.

Students and learning

Somewhat paradoxically, students' low levels of ability in using computers and the Internet for learning was also a significant concern of teachers. They were concerned about whether students had the research skills and strategies needed for learning on the web, and they were worried that students did not have the skills necessary to search effectively. Teachers remarked that "students have to spend quite a lot of time to search for useful information," they have to learn "to be selective when they are searching on the net," and they may "feel frustrated in the searching process since there are lots of things on the net." Warschauer (1997) states that Internet activities can result in various complexities that are different from the traditional classroom and that students may not have the prerequisite computer skills necessary for such activities.

Another issue mentioned by the teachers was maintaining students' attention when using computers. The computer was perceived as a competitor for students' attention, the competition being between the educational programs and the computer programs: "Students may be easily distracted from studies as there are lots of entertainments."

These beliefs underlie a misconception about the learning opportunities offered by computers and their potential outcomes. Such beliefs indicate that teachers may have overlooked one of the most essential and important benefits that online communication offers to learning, that is, the opportunities for meaningful interaction and communication. Students can interact with each other without the restrictions of place and time. They can learn collaboratively and actively online with other learners, by exchanging ideas, negotiating meanings, sharing experience, and constructing knowledge.

In addition to more specific comments about the perceived negative effects of using computers on the development of students' language skills, a few groups of teachers had similar concerns about the effectiveness of technology for teaching and learning English in general. They commented, "it is doubtful whether students get more exposure to English" and "whether students' exposure to English is actually enhanced through the use of IT." One teacher added, "language learning is more complex; it may not be the best way to use computers to teach English."

Teachers and teaching

Controlling and monitoring students' access to websites and participation on the web were also common concerns of teachers. One of the control issues relates to students accessing inappropriate Internet sites. Teachers pointed out that there is "no control on negative information"; there is a need for teachers to prevent students from "accessing dangerous websites."

Many teachers worried that "students may navigate other parts of the web by themselves ... not focusing on what teachers ask them to do." Teachers believed that "students need(ed) guidance and have to be monitored to stay on task." The teacher's role was to ensure that students stayed on task by choosing sites for them, guiding them to navigate, and monitoring their online activities.

Such comments suggest a strong teacher preference for controlling and monitoring. These points are related to more fundamental beliefs about the roles of teachers and their students, and about concepts of learning and teaching. As Becker (1991) argues, the use of technology "will require teachers and administrators to modify their concepts of appropriate and inappropriate teaching behaviors, to reprioritize the value of different types of instructional content, and to change habits and assumptions that guide their classroom and school management strategies" (p. 8).

Access and equipment

The reliability of equipment and access was another critical issue for teachers. Many mentioned their worries about technical problems. Some stated, "technical problems happen usually unexpectedly," "equipment failure may hinder the lesson," and some other "unexpected technical problems may affect the flow of the lesson." A group of teachers noted that it would be embarrassing if they were unable to deal with technical problems, and others suggested "slow access will create dead air."

Unpredictable occurrences such as computer breakdown and problems with Internet connection have created fears and frustrations for teachers. Clearly, there is a need for certainty and security in the process of teaching with computers.

Other problems noted by the teachers include logistic and practical ones. In order to use the MMLC, teachers have to check out the availability of the room. English classes have to share access with other subjects. However, as Singhal (1997) found, for the most part, computers in schools are used for business or computer science classes; thus, English teachers might not have many opportunities to use the special room. A teacher said in the focus group discussion, "more MMLC is needed, as there are a lot of classes."

A major practical problem mentioned by teachers was the time needed to gain access to the MMLC and to set up the equipment and Internet connections for each forty-minute lesson. The time factor was perceived as a significant obstacle impacting on the willingness of teachers to use technology. These apparently deep-seated perceptions about IT and its advantages and disadvantages raise the question of how effective short in-service programs can be skilling teachers and changing their underlying beliefs. The next section examines the effectiveness of the in-service program.

EFFECTIVENESS OF THE IN-SERVICE PROGRAM

In this section, data from the end-course questionnaire and focus group interviews are presented and discussed, to show the nature of the changes over the course of the in-service program in the teachers' attitudes, beliefs, and views of the use of IT in teaching, their perceived IT skill level, their plans in using IT, and the effectiveness of the program in helping teachers in implementing IT in teaching.

Table 10.2 Course usefulness

Very useful	Useful	Barely useful	Not useful
2	9	1	0
17%	75%	8%	0%

n=12

Table 10.2 shows the overall rating of the usefulness of the course. Ninety-two percent of the teachers rated the overall usefulness of the course as "very useful" or "useful." The teachers were very positive about the course in all the aspects: Objectives, Coverage, Content, Progress, Duration, Preparation, Materials, Website, Instructor Presentation, Practical Work, and Assistance. This demonstrates that the course was very successful in meeting teachers' expectations and satisfying their needs.

In the end-course questionnaire, teachers commented that they liked the course because it was practical, they were made aware of some useful resources, and they were happy that they were able to create some online interactive activities. In addition, they liked the opportunities they had for group discussion and interaction. They also appreciated that the course was taught "by English teachers to English teachers," and they were satisfied that the instructors were "very approachable, helpful and professional." They commented that the course had improved their confidence in using IT, one teacher summing up the overall change in attitude: "IT can be friendly."

In both the end-course questionnaire and focus group discussions, teachers indicated that the course had helped them professionally in a number of ways. They felt better equipped to select and use relevant resources for teaching, discover more websites and activities on teaching the English language, create online interactive activities, develop new ideas for using IT for teaching (for example, ideas on marking online), finding out about designing materials with IT that suited the needs of students, and using IT resources. In addition, the teachers stated that they were more "positive," "interested," and "confident" in the use of IT after the course.

A comparison of the teachers' perceived IT skill levels before and after the course is shown in Figure 10.4. Thirteen percent of the teachers felt anxious about their IT skills before the course, but no teacher felt anxious at the end of the course. Moreover, the percentage of teachers feeling competent (rated "4") about their IT skills increased from thirteen percent to seventeen percent. Although the data show that teachers' perceived IT competence did not increase greatly as a result of the course, the overall competence level of this group of teachers generally improved. This result is to be expected, as the short duration of the course meant teachers did not have time to explore and practice what they had learned.

		1	2	3	4	5	Average
Pre-course	N=15	2	3	8	2	0	3
		13%	20%	53%	13%	0%	
End-course	N=12	0	4	6	2	0	2.4
		0%	33%	50%	17%	0%	

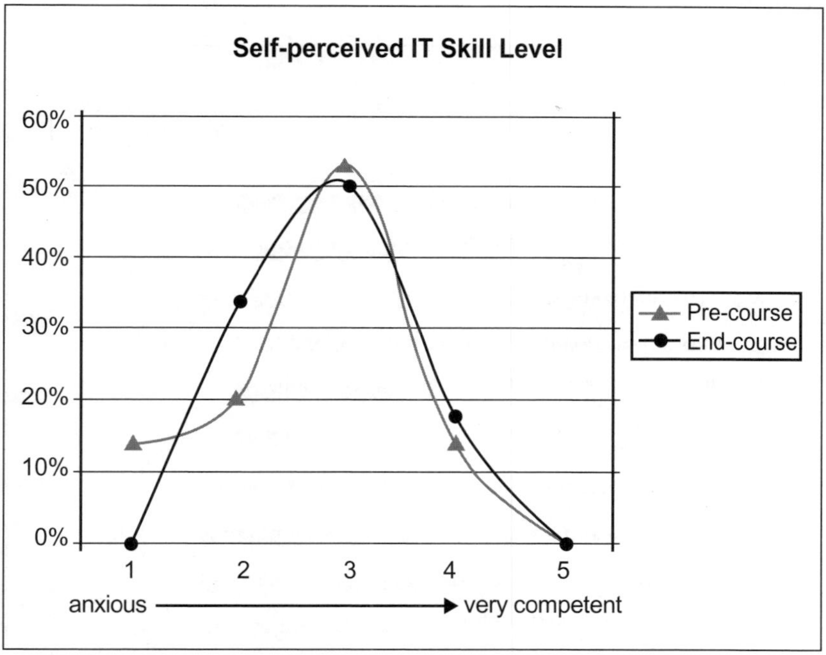

Figure 10.4 Perceived IT skills before and after the course

In fact, when we compare the average rate of the teachers' perceived IT skill levels before and after the course, the figures show that the rate was slightly higher for the pre-course (3) than for the end-course (2.4). This is not so surprising, because the teachers certainly had a better idea about IT and its use after the course, so they might have become more aware of their own lack of knowledge or skills. However, this is not necessarily negative, as indicated by their comments above. The comments suggest that they were motivated to know more and learn more, and that the course had aroused their interest.

The teachers' views and attitudes about the use of IT at the end of the course are shown in Figure 10.5. The results are very positive, no item receiving a negative rating. The items rated most positively were those to do with the teachers' interest in IT and use of IT. Those related to skill competence were rated comparatively lower. As the discussion of results in the previous section revealed, this course was very short, teachers being exposed to the use of IT in English teaching as well as some IT skills and information but given no time to apply and practice what they had learned, so they were not so sure about the applications and their own skill development.

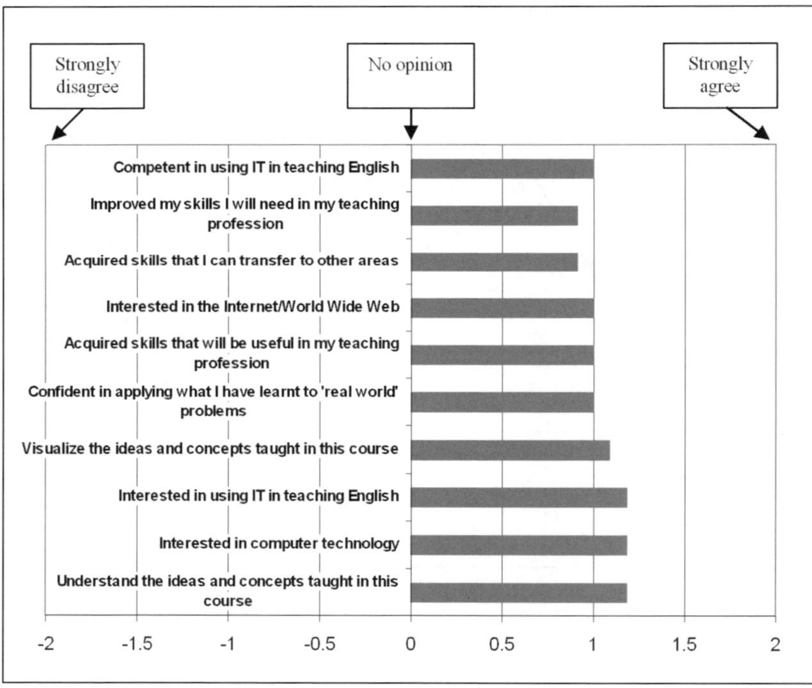

Figure 10.5 Course effectiveness

A comparison of the average self-evaluation of different skills by the teachers before and after the course is shown in Figure 10.6. Teachers improved their skills in all the areas, the most significant increase in their perceived ability being to *create interactive quizzes and activities on the web*, *mark online*, *create or edit a website*, and *interact via time-delayed communication*. Between forty and fifty-seven percent of the teachers indicated that they had no knowledge or ability in these four areas before the course. In contrast, fifty-eight to seventy-three percent of the teachers rated themselves as average in these four skills areas at the end of the course.

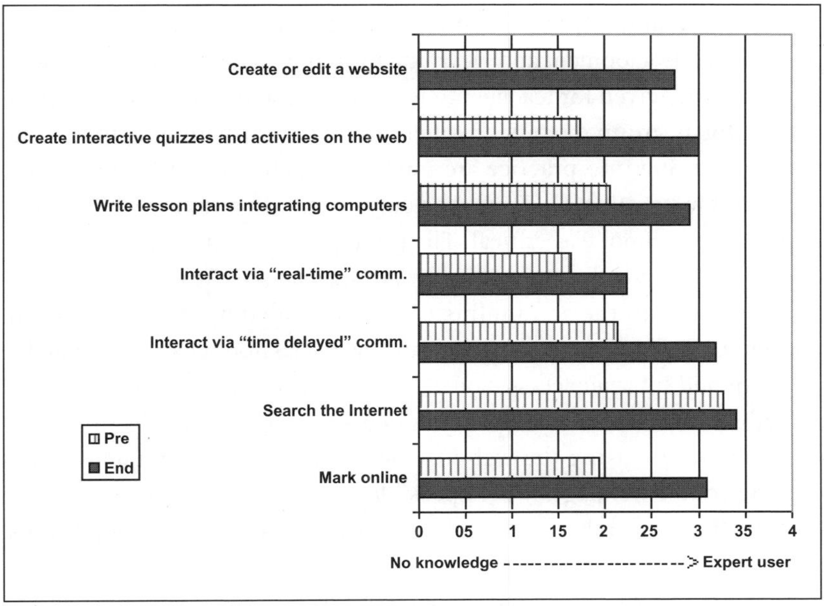

Figure 10.6 Ability in using computers before and after the course

Teachers commented that the course gave them a basic introduction to the integration of IT in English language teaching and learning and they felt that they knew more about IT and the opportunities it could provide at the end of the training sessions. Although they did not have a very clear idea of what changes they would instigate, most were open to the idea of trying out some ideas and skills learned in the course. However, three teachers said they still would not use IT or were hesitant to use it because of time constraints. Another teacher pointed out that she was "still scared to do so and felt incompetent." These comments support the findings discussed previously about the importance of time as a factor in teachers' uptake of IT in the classroom.

Teachers emphasized building up resources and collaboration. In the focus group discussions, ideas like "collaboration and sharing teaching materials, centralize IT resources," "build up a pool of resources," and "dig into exchange materials" came up most. Some teachers also expected that there would be "a lot more commercial materials" in the near future. Besides working collaboratively on creating and sharing teaching resources and materials, teachers saw the importance of making "connections with other schools and even university to get access to materials." Some suggested the use of bulletin boards for communication with colleagues and other schools.

Another major plan for most of the teachers was continuing professional development and "learning by doing." They really appreciated "opportunities given for teachers to learn, to upgrade and to try." Teachers in both focus group discussions mentioned that attending courses and putting learning into practice are important factors contributing to the successful integration of IT into the curriculum. In addition, they stated support from school was critical. The principal's enthusiasm in organizing the course was greatly appreciated by all the teachers. Teachers were also very pleased with the good infrastructure provided by the school. Each teacher was given a laptop computer, and every classroom had a multimedia projector and Internet access. Some specific plans were also mentioned by individual teachers, such as asking students to submit work electronically, marking online, using the web for communication with students, and creating course webpages where possible.

The findings of this section of the study have significant implications for professional development and teacher education programs. These implications are discussed in the following section.

IMPLICATIONS FOR TEACHER EDUCATION

The rapid spread of IT technologies places unusual stresses on teachers. The new technologies often seem incompatible with their established working habits, creating uncertainties and even threats to the teachers' own professional status (see Davison, this volume, for further discussion of this point). Such technological challenge highlights the importance of teachers' beliefs and attitudes toward technologies and their integration into teaching and learning.

Engendering teacher change requires profound changes in educational beliefs, attitudes, and practices of teachers (Kennedy 1987). Such changes affect teachers' "professional self-definition" (Fullan 1982, 115). Changes also require teachers to expand and alter their perceptions (Freeman 1989) and to learn new ways of thinking and practicing in teaching, and at the same time learning new skills, knowledge, and attitudes (Fullan 2001). As Fullan (1993, 105) argues:

> Teacher education still has the honor of being simultaneously the worst problem and the best solution in education.

The emergence of new information technologies provides the opportunity and the necessity for change. Teacher education plays an essential role in this process of change, in that it represents "an attempt to implement some form of change" (Kennedy 1987, 163). Its role is to activate the teacher to engage in "the process of reflection, examination, and change" and thus to improve performance and experience personal and professional growth (Freeman 1982, 21). Many studies have shown that teacher education is closely related to successful implementation of innovation (Parish and Arrends 1983; Fullan and Hargreaves 1992; Parr 1999), provided the following qualities for teacher support and professional development are fulfilled.

Ongoing, systematic, comprehensive, and long term

Any innovation takes time to implement and requires extensive long-term professional development. Ongoing teacher education is essential for teachers to continually and critically reflect on their own teaching as well as their own learning. A number of researchers (Lange 1990; Nunan 1989) emphasize the importance of an open-ended, experiential, and developmental model of teacher education. Hopkins et al. (1994) point out that teacher development must take place in the workshop context and in the workshop. That means teachers are presented with new ideas, new practices, and new tools; they then learn to handle them and experiment with them by transferring the skills and ideas to their own teaching context, to identify problems and find solutions. This ongoing development needs the commitment of time to understand new ideas and new roles and their implications, to develop the appropriate skills in trying out and adapting new ideas in the teaching context. It also requires a systematic process of learning and development, application and implementation, evaluation and reflection. Brindley and Hood (1990) stress the importance of continuing professional development:

> If teachers are being asked to change some aspect of their classroom behavior they need professional development activities which enable them at the same time to use an innovation and to work through the implications of the change with colleagues. This requires an ongoing program in which teachers commit themselves to classroom action followed by reflection and theory as necessary. (p. 245)

Information technologies provide the potential for teacher education to be moved beyond centralized training institutions into the teachers' own working environment. In addition, they open up the opportunity for continuing and long-term teacher education. Ongoing learning and development is especially essential for integrating technologies into education; short-term refresher programs only show teachers the new opportunities technologies provided but do not help them to cash in on the opportunities. As Langone et al. (1998) suggest, ongoing coaching at school for several years after taking a course might be a way to support teachers in technology implementation in specific contexts and bringing effective change.

Collaboration, communication, and support

Teacher education also needs to be geared towards building a network between teachers with whom they can share experiences and good practices in using IT in language teaching and collaborating to construct knowledge and develop resources. Teachers should be encouraged to take an active role in the innovative process of integrating information technology into teaching. Information technology can play a crucial role in facilitating greater communication, collaboration, understanding, and reflection among different teachers. One of the most critical developmental factors for teachers is an environment in which they can share their effort and achievements and overcome any feelings of isolation. Teachers need to consider and develop ways in which they can collaborate with others outside their school, within their school, with their students, and potentially even with parents. English teachers can work together or even collaborate with teachers of other subjects to develop IT skills and integrate IT across the curriculum. This type of collaboration reduces everyone's workload and allows teachers to remain more motivated and to learn from each other. This partnership between different participants reinforces learning and enhances professional growth.

This kind of collaborative learning and support is also important for the sharing of ongoing experiences, uncertainties, successes, problems, difficulties, materials, and resources in a supportive environment. Teachers' attitudes towards an innovation are often influenced by how worthwhile and important others within their working environment perceive the innovation to be (Kennedy 1996). Supportive environments provided by the school administration facilitate the acceptance of an innovation and help reduce the anxiety and difficulty of learning new skills (Hopkins et al. 1994).

Another mode of collaboration is for teachers to sign up for courses together or to seek out collaborators within a course that they are taking. However, one of the greatest shortcomings of most in-service courses is that they do not offer any long-term potential for collaboration. There are, however, different opportunities for teachers to meet and collaborate and learn with others using IT. In order to stay abreast of current issues and to tap into a larger community, teachers can participate in mailing lists or use discussion boards to meet and exchange ideas with other teachers and educators, locally and internationally.

Contextualized: Taking into account teachers' knowledge, experience, needs, attitudes, and beliefs

One of the greatest potentials of information technologies in teacher education is that flexibility in facilitating the moving away from generic teacher training programs to allow the development of specific teacher education models that are sensitive to the various needs, experience, learning styles, knowledge and background, working culture and situations of teachers. The kinds of more individualized and adaptive approaches to teacher education allowed by information technologies are better in addressing the critical issues of attitudes, motivation, and beliefs for achieving successful and effective innovation.

A number of researchers (Dalin and Rolff 1993; Fullan 1993; Sarason 1990; Nunan 1992) identify teacher attitudes as context-specific and the specific context as critical in influencing teachers' attitude change. Kennedy (1987) asserts the importance of context-specific beliefs in ELT innovations:

> Views held on theories of language teaching and learning and views on the educational process and what happens or should happen in classrooms between the teacher and students are ultimately context specific and derived from the culture of the society in which learning takes place. (p. 166)

Creating change agents: Teacher-initiated and teacher-directed change

The one-off mass training sessions in which teachers are bombarded with new theories, concepts, skills, and teaching approaches are usually ineffective in bringing sustainable changes. Teacher education should involve long-term development that results from reflection and critical awareness (Freeman 1989; Richards 2001). It should help teachers to become lifelong learners and take an active role in developing their ability to raise questions, to reflect upon and evaluate their own practice. Technology-mediated open learning environments stimulate awareness, reflection, and problem solving. Teachers are free to explore the potentials and constraints of adopting the innovation, to define, clarify, apply, reflect, and modify their own knowledge, perceptions, attitudes, beliefs and practices to address their needs, concerns, and problems in their own working situations. Thus, information technologies take an important role in developing teachers as the change agents of innovations.

Balancing pedagogy and technology

Comprehensive professional development for technology should include not only the development of technical skills and knowledge but also strategies for technology-enhanced teaching and learning and for classroom management and practice. It should help teachers develop theoretical and pedagogical competence in applying technology in their own practicing environment and integrating IT into the larger curricular framework, so that technology will become a powerful tool that is truly useful to teachers in making the pedagogical shift from being the "sage on the stage" to mentors who are "guides on the side." Teacher education should demonstrate this balance between pedagogy and technology to prepare teachers for their new roles in technology-mediated environments. Teacher education then models the approaches and practices teachers are urged to adopt in integrating technologies in teaching and learning.

CONCLUSION

Although the research reported here is a small-scale, preliminary study, it provides some useful insights into in-service teachers' beliefs about technology and its use. It shows that teachers are aware of the benefits of IT in increasing the motivation of students and the quality of their learning, but they are also very concerned that producing IT-based materials takes a lot of time. Because of a heavy workload and high teacher-student ratio in

Hong Kong schools, it is difficult for English teachers to find the time required to explore the potential of IT, to consolidate and extend their skills, or to try out new teaching ideas. Furthermore, the study showed that teachers are generally worried about their perceived lack of IT skills and knowledge about the use of IT in teaching English, and they do not know where to find suitable computer-assisted language learning materials. Thus, they really value training opportunities that suit their needs.

The study helps us to understand teachers' perceived IT skills and knowledge, and their need for support. The teachers' description of the major difficulties and obstacles to the implementation of IT in their schools reflects their perceptions of the relative importance of the different types of challenges they face. The study provides a basic framework for larger scale study of a similar type in revealing the beliefs and values that support or hinder teachers' effective use of IT in teaching. In addition, it shows the importance of understanding what change and innovation means for teachers, and teachers' beliefs and reactions to change. It also indicates that professional development is critical for helping teachers to see the opportunities with IT and to figure out how IT can help them in enhancing the quality of teaching and learning. As Cuban (1997) states, the kind of professional development needed is to "involve teachers deeply and continuously in on-site learning; hang in with them as the inevitable squalls of turmoil blow and recede; and finally, have patience, for such changes in belief and practice will take years" (p. xiv).

In addition to revealing teachers' beliefs about the opportunities and challenges in using technology in teaching, this study emphasizes the importance of teacher professional development in the success of the innovation, and the research implications suggest some principles in supporting teachers in the implementation of IT in teaching.

NOTE

1. A large-scale longitudinal study that aims at exploring the beliefs of English teachers about the implementation of IT in the curriculum and the impact of technological innovations on pedagogical practices and teachers' belief system is in progress. Contact the authors for further details about the study and the in-service training program.

REFERENCES

Adams, D. (1985). *A look back — A look ahead. Computer and teacher training: A practical guide*. New York: Haworth Press.

Becker, H. (1991, May). When powerful tools meet conventional beliefs and institutional constraints. *The Computing Teacher*.

Brindley, G. and Hood, S. (1990). Curriculum innovation in adult ESL. In G. Brindley (ed.) *The second language curriculum in action* (pp. 232–47). Sydney: NCELTR.

Byron, I. (1997). Information and communication technologies in education. *Educational Innovation and Information, No. 91*.

Cooper, R. (1989). *Language planning and social change*. Cambridge: Cambridge University Press.

Cuban, L. (1997). Foreword. In J. Sandholtz, C. Ringstaff, and D. Dwyer (eds.) *Teaching with technology: Creating student-centered classrooms* (pp. xi–xiv). New York: Teachers College Press.

Curriculum Development Council. (1999). *Syllabuses for secondary schools: English language (Secondary 1–5)*. Hong Kong: The Education Department.

Dalin, P. and Rolff, H. (1993). *Changing the school culture*. London: Cassell.

Drucker, P. (1992). *Managing for the future: The 1990s and beyond*. New York: Truman Talley Books.

Dunkel, P. (1987). Computer-assisted instruction (CAI) and computer-assisted language learning (CALL): Past dilemmas and future prospects for audible CALL. *Modern Language Journal, 71*, 250–60.

Dwyer, D. (1994). Apple classroom of tomorrow: What we've learned. *Educational Leadership, 51*, 4–10.

Education and Manpower Bureau. (1998). *Information technology for quality education: Five-year strategy 1998/99 to 2002/03*. Hong Kong: Hong Kong Special Administrative Region Government.

Education Commission. (2000). *Learning for life, learning through life: Reform proposals for the education system in Hong Kong*.

Ehrmann, S.C. and Zúñiga. R.E. (1997). *The flashlight evaluation handbook*. Washington, DC: The Teaching, Learning, and Technology Group.

Fang, Z. (1996). A review of research on teacher beliefs and practices. *Educational Research, 38*(1), 47–65.

Freeman, D. (1982). Observing teachers: Three approaches to in-service training and development. *TESOL Quarterly, 16*(1), 21–8.

Freeman, D. (1989). Teacher training, development, and decision making: A model of teaching and related strategies for language teacher education. *TESOL Quarterly, 23*(1), 27–45.

Fullan, M. (1982). *The meaning of education change.* New York: Teachers College Press.

Fullan, M. (1993). *Change forces: Probing the depths of educational reform.* New York: Falmer Press.

Fullan, M. and Hargreaves, A. (1992). Teacher development and educational change. In M. Fullan and A. Hargreaves (eds.) *Teacher development and educational change* (pp. 1–9). London: Falmer Press.

Fullan, M. (2001). *The new meaning of education change (3rd ed.).* New York: Teachers College Press.

Garrett, N. (1991). Technology in the service of language learning: Trends and issues. *The Modern Language Journal, 75*(1), 74–101.

Hannafin, R. and Savenye, W. (1993). Technology in the classroom: The teacher's new role and resistance to it. *Educational Technology*, 26–31.

Hanson-Smith, E. (ed.) (2000). *Technology-enhanced learning environments.* Alexandria, VA: TESOL.

Henrichsen, L. (1989). *Diffusion of innovations in English language teaching: The ELEC effort in Japan, 1956–1968.* New York: Greenwood Press.

Hopkins, D., Ainscow, M., and West, M. (1994). *School improvement in an era of change.* London: Cassell.

Kennedy, C. (1987). Innovating for a change: Teacher development and innovation. *ELT Journal 41*(3), 163–70.

Kennedy, C. (1988). Evaluation of the management of change in ELT projects. *Applied Linguistics, 9*(4), 329–42.

Kennedy, C. (1996). Teacher roles in curriculum reform. *English Language Teacher Education and Development* 2(2), 52–65.

Knupfer, N. (1993). Teachers and educational computing: Changing roles and changing pedagogy. In R. Muffoletter and N. Knupfer (eds.) *Computers in education: Social, political, and historical perspectives* (pp. 163–80). Cresskill, NJ: Hampton Press.

Lange, D. (1990). A blueprint for teacher development. In J. Richards and D. Nunan (eds.) *Second language teacher education* (pp. 245–68). Cambridge: Cambridge University Press.

Langone, C., Wissick, C., Langone, J., and Ross, G. (1998). A study of graduates of a technology teacher preparation program. *Journal of Technology and Teacher Education, 6*(4), 283–302.

Law, N. (2000). Learning in a new era — or an expensive trip to a "technology rich" educational wasteland? Hong Kong: Centre for IT in School and Teacher Education, The University of Hong Kong.

Marcinkiewicz, H. (1993). Computers and teachers: Factors influencing computer use in the classroom. *Journal of Research on Computing in Education,* 26, 220–37.

Markee, N. (1986). Towards an appropriate technology model of communicative course design. *English for Specific Purposes, 5*, 161–72.

Markee, N. (1993). The diffusion of innovation in language teaching. *Annual Review of Applied Linguistics 13*, 229–43.

Markee, N. (1997). *Managing curricular innovation*. Cambridge: Cambridge University Press.

Mehlinger, H. (1995). *School reform in the information age*. Bloomington, IN: Center of Excellence in Education.

Nunan, D. (1989). Second-language teacher education: Present trends and future prospects. In C. Candlin and T. McNamara (eds.) *Language, learning and community* (pp. 143–54). Sydney: NCELTR.

Nunan, D. (1992). The teacher as decision-maker. In J. Flowerdew, M. Brock, and S. Hsia (eds.) *Perspectives on second language teacher education* (pp. 135–65). Hong Kong: City Polytechnic of Hong Kong.

Pajares, M. (1992). Teachers' beliefs and educational research: Cleaning up a messy construct. *Review of Educational Research, 62*(3), 307–32.

Parish, R. and Arrends, R. (1983). Why innovative programs are discontinued. *Educational Leadership 40*(4), 62–5.

Parr, J. (1999). Extending educational computing: A case of extensive teacher development and support. *Journal of Research on Computing in Education, 31* (3), 280–91.

Pennington, M.C. (ed.) (1996). *The power of CALL*. Houston, TX: Athelstan Publications.

Pusack, J. and Otto, S. (1997). Taking control of multimedia. In M. Bush and R. Terry (eds.) *Technology-enhanced language learning* (pp. 1–46). Lincolnwood, IL: National Textbook Company.

Richards, J. (2001). Postscript: The ideology of TESOL. In R. Carter and D. Nunan (eds.) *The Cambridge guide to teaching English to speakers of other languages* (pp. 213–7). Cambridge: Cambridge University Press.

Richards, J. and Lockhart C. (1996). *Reflective teaching in second language classrooms*. Cambridge: Cambridge University Press.

Ridgway, J. (1997). Vygotsky, informatics capability, and professional development. In D. Passey and B. Samways (eds.) *Information technology: Supporting change through teacher education* (pp. 3–19). London: Chapman and Hall.

Robinson, B. (1997). Getting ready to change: The place of change theory in the information technology education of teachers. In D. Passey and B. Samways (eds.) *Information technology: Supporting change through teacher education* (pp. 40–5). London: Chapman and Hall.

Rogers, E. (1995). *The diffusion of innovations* (4th ed.) New York: The Free Press.

Sarason, S. (1990). *The predictable failure of educational reform*. San Francisco: Jossey-Bass.

Scrimshaw, P. (1995). Computers and the teacher's role. In B. Somekh and N. Davis (eds.) *Using information technology effectively in teaching and learning: Studies in pre-service and in-service teacher education*. London and New York: Routledge.

Singhal, M. (1997). The Internet and foreign language education: Benefits and challenges. *The Internet TESL Journal*, *3*(6). Retrieved 12 June 2002 from http://www.aitech.ac.jp/~iteslj/Articles/Singhal-Internet.html.

Stern, H. (1983). *Fundamental concepts of language teaching*. Oxford: Oxford University Press.

Stern, H. (1992). *Issues and options in language teaching*. Oxford: Oxford University Press.

Stoller, F. L. (1994). The diffusion of innovations in intensive ESL programs. *Applied Linguistics 15*(3), 300–27.

Terrell, S., Dringus, L., and Rendukuc, P. (1995). *A transitional model for the introduction of technology*. (ERIC Document Reproduction Service No. ED 386 171).

Warschauer, M. (1997). The Internet for English teaching: Guidelines for teachers. *The Internet TESL Journal*, *3*, 10. Retrieved 12 June 2002 from http://www.aitech.ac.jp/~iteslj/Articles/Warschauer-Internet.html.

White, R. (1988). *The ELT curriculum: Design, innovation and management*. Oxford: Blackwell.

Zaltman, G. and Duncan, R. (1977). *Strategies for planned change*. New York: John Wiley and Sons.

Zammit, S. (1992). Factors facilitating or hindering the use of computers in schools. *Educational Research, 34*, 57–66.

11

The *TeleNex* Pedagogic Primary Grammar: Theoretical and Navigational Issues

Quentin Allan

INTRODUCTION

The final decade of the twentieth century witnessed some serious re-examination of educational goals in Hong Kong, particularly with regard to language standards. Because of the rapid expansion of tertiary education in the 1990s, there has been a concern in the Hong Kong community that English language standards are falling, and thus English teaching in general has come in for particularly close scrutiny. Research undertaken in the mid-1990s indicated that a significant majority of English language teachers were neither professionally nor subject trained (Lock and Tsui 1996). Given the restricted domains in which English is used as a lingua franca in Hong Kong, many teachers also had problems with proficiency in English. There has also been a growing awareness that teacher support is a career-long need; teachers are likely to function more effectively if they have a sense of belonging to a cohesive, professional community. What was needed, then, was a centrally co-ordinated mechanism that would have the capacity to provide continuous professional development and support throughout one's entire teaching career. To fill this need, the *Teachers of English Language Education Centre* (TELEC), was established in 1993 under the aegis of the Department of Curriculum and Educational Studies at the University of Hong Kong. TELEC's main objectives are to provide appropriate resources for the professional development of English language teachers and to encourage collegiality and mutual support, i.e., to promote a sense of professional community.

Given the sheer number of English teachers dispersed throughout Hong Kong and their limited time due to extremely heavy workloads, the Internet suggested itself as a likely medium for professional development and support. *TeleNex* was therefore established in order to bring teachers together in a virtual community and to make available pedagogically sound teaching materials and an array of professional development resources aimed at improving their English language knowledge base. *TeleNex* comprises both a primary and secondary website and is available to all registered teachers in Hong Kong. The secondary website was established from 1993–98 and, in the course of its five-year gestation, many lessons were learned about website architecture, navigation, interface design, and the dynamics of Computer Mediated Communication (CMC). These lessons have informed the development of the primary website that was developed from 1998–2002.

TeleNex, being web-based, is able to exploit a rapidly developing medium, and transcends traditional financial, temporal, and geographical constraints. It draws on developments and experience with *asynchronous learning networks* (ALNs), which are web-based learning venues that emphasize people-to-people communication in conjunction with traditional and/or multimedia learning tools. ALNs enable people to learn and develop professionally through self-study and in interaction with other interested parties through email and/or threaded conferencing systems. Bourne (1998) has traced the development of such networks, noting their impact in many educational arenas including on-campus education, off-campus education, and continuing education. For further background relating specifically to the development of *TeleNex*, including a full discussion of TELEC's aims and objectives, see Wu and Tsui (1997).

This chapter begins by briefly outlining the main features of *TeleNex*. The main focus is on the theoretical and navigational issues associated with the development of one component in particular, a hypertext pedagogic grammar (*PrimeGram*), which has been designed specifically for primary teachers. There have been a number of special challenges in developing *PrimeGram*. These include the particular requirements of the target readership, adoption of a functional model of language, and the introduction of a major curriculum reform (Target Oriented Curriculum). On the technical side, the use of hypertext and multimedia pose a number of special problems: the requirement that navigation be simple, intuitive, and consistent proving to be one of the greatest challenges. The mechanisms that are currently in place have evolved through eight years of trialing and adaptation in response to users' feedback. The final section of the chapter touches on the question of sustaining the network from the dual perspectives of electronic database management and continued user involvement.

THE COMPONENTS OF *TeleNex*

The *TeleNex* website can be accessed free of charge by registered teachers twenty-four hours a day at the point of need. Registration is a straightforward procedure over the Internet, involving confirmation that the applicant is a bona fide teacher in a local school. A sample of files from all databases, plus sample concordances from *PatternFinder* and sample messages from the Conference Corners can be accessed at http://www.telenex.hku.hk. Figure 11.1 shows the opening screen of the website.

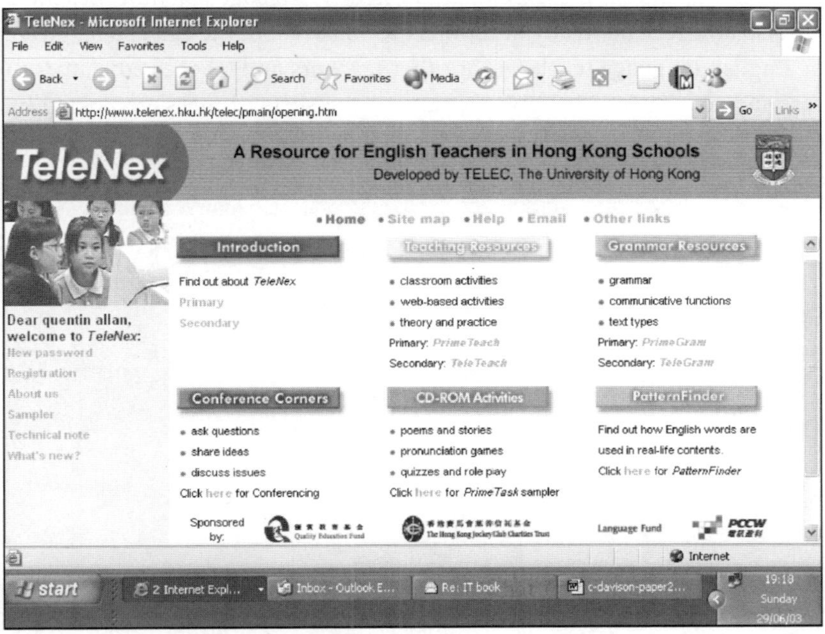

Figure 11.1 TeleNex opening screen

The major components of the *TeleNex* network available to teachers are summarized in Table 11.1.

Table 11.1 Summary of the available TeleNex components

Resources for Secondary Teachers	Resources for Primary Teachers
TeleTeach (teaching materials)	*PrimeTeach* (teaching materials) *PrimeTask* (interactive multimedia tasks)
TeleGram (a pedagogic grammar)	*PrimeGram* (a pedagogic grammar)
PatternFinder (a concordancer and corpora)	
Conference Corners	

One of the most important components of the *TeleNex* network is the *Conference Corners*, an interactive, threaded discussion facility that is shared by both primary and secondary teachers. This component comprises a collection of theme-based corners through which teachers are encouraged to share, in addition to classroom teaching materials, their ideas, experiences, and problems related to English language teaching. Teachers are also able to send in any questions they have about the English language or English language teaching for other teachers and teacher educators to respond to. Figure 11.2 gives a taste of the sorts of message posted by teachers in the Primary Teaching Corner.

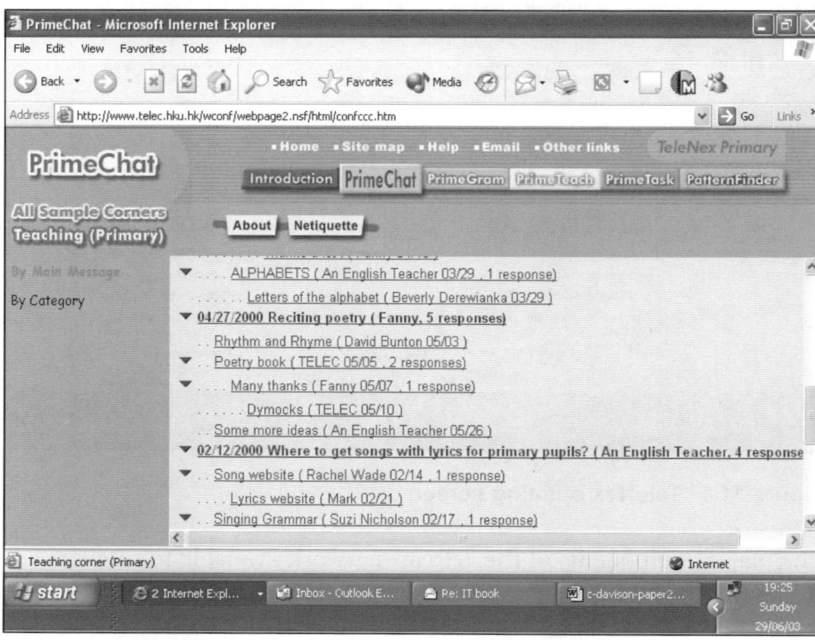

Figure 11.2 Sample messages from the Primary Teaching Corner

As participation in the Conference Corners is restricted to registered teachers in local schools, the facility provides a private space for teachers to air their concerns and grievances candidly and without fear of censure from parents and members of the public. In 2002, it was decided to open the *TeleNex* network to staff in the Curriculum Development Institute (CDI) so that they can better help teachers engaged in school-based curriculum development and teaching. Both primary and secondary teachers are encouraged to engage in discussions related to teaching and language issues, and they may do so using their own name or the anonymous *An English Teacher*. For further information about various aspects of conferencing, see Tsui and Ki (1996), Wu and Sengupta (1998), Tyrwhitt-Drake (1999), and Tsui and Bunton (2000).

The *PatternFinder* component (a web concordancer) has been provided to enable teachers to conduct their own searches into collections of native speaker and learner corpora. Concordance lines are typically used to illustrate collocational and colligational patterns of standard and non-standard usage. For example, in response to a message querying whether *everyday* written as one word can be used as an adverbial phrase, a short concordance is provided to show that in its one-word form, it is typically used only as a classifying adjective. Figure 11.3 shows the opening screen of *PatternFinder* with a sample concordance of *homeworks* to illustrate a typical search result from a corpus of student writing.

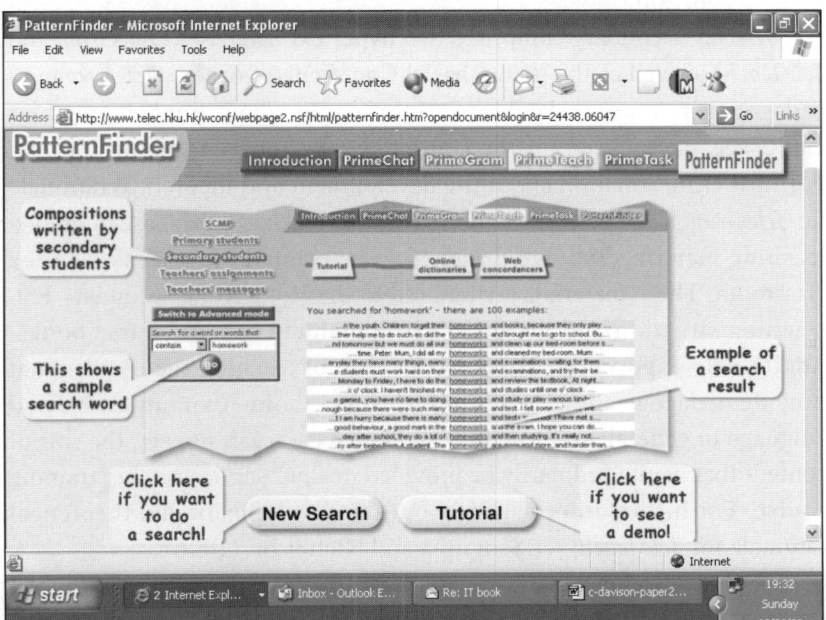

Figure 11.3 The opening screen of PatternFinder

Teachers' reactions to the use of concordance lines in Conference Corners messages have been very positive, and a growing number of teachers have expressed interest in being able to use a concordancer themselves. It is becoming increasingly apparent that access to various corpora has a unique potential to raise teachers' awareness of patterning and usage within specific genres and texts as well as across the language as a whole. Berry (1994, 107) notes that through the process of analyzing data in the form of concordance lines and frequency lists, teachers can develop confidence in their intuitions as well as discover their limitations. Furthermore, the systematic analysis of concordance lines lends itself to an inductive approach and provides a focus for the development of analytical skills following Johns' (1988) notion of the *teacher as researcher* and in keeping with the current trend towards teacher-initiated action research. The data-driven learning approach is, moreover, inherently motivating: Teachers can conduct tailor-made searches relevant to their own interests and preoccupations. By using a discovery-learning approach, the teacher comes to "own" the knowledge gained: and, as the findings from any given search will invariably suggest other avenues for exploration, the process of investigation and analysis will be never-ending. Teacher educators also feel that interacting with corpora will help teachers to improve their own command of the English language. For a more extended discussion, see Allan (1998, 2003).

TeleNex Secondary comprises two hypertext databases (*TeleGram* and *TeleTeach*) as well as the theme-based Conference Corners. *TeleGram* is a pedagogic grammar, a database of information about English grammar and usage, customized for the Hong Kong secondary teaching context. For further information about the development and theoretical rationale for *TeleGram,* see Lock and Tsui (1996). *TeleTeach* is a database of graded teaching materials that are designed to be printed out and used in the classroom. These materials are tailored to the Hong Kong secondary ESL teaching situation and designed to supplement school course books. Additional support is available in the form of teaching theory files that provide pedagogically oriented information about communication and language in general, and teaching and language teaching, i.e., the sort of content that would ordinarily be provided in a pre-service teacher training course. For further information about the development and theoretical rationale for *TeleTeach,* see Sengupta and Nicholson (1996).

Through extended interaction with these *TeleNex* resources, it is hoped that teachers will improve their understanding of and proficiency in

English, develop a wider repertoire of teaching skills, and come to a deeper understanding of theoretical principles that will help them to make better informed decisions about their classroom practice. The *TeleNex* network is underpinned by integration among support for teachers, research into the felt needs of teachers, and ongoing development of the resources. The questions and responses that teachers send in via the Conference Corners serve as input to TELEC research projects and database development. The research output feeds into the professional advice and support provided to teachers through Conference Corners and databases. Use of these resources also generates feedback (solicited and ad hoc) that is vital for ongoing development.

The next part of this chapter elaborates on the development of the primary website, with particular reference to the pedagogic grammar, *PrimeGram*.

DEVELOPMENT OF THE PRIMARY PROJECT

Development of the primary project began at a time when the Curriculum Development Council (CDC), with the endorsement of the Education Department, was promoting a major curriculum reform, the Target Oriented Curriculum (TOC) that entailed a shift in emphasis from structural accuracy to development of communicative competence, while retaining a strong focus on grammar. The TOC promotes a task-based approach and is designed to give "a clearer direction to teaching, learning, and assessment by setting common learning targets and objectives for all learners for each Key Stage" (CDC 1997, 133).

Assessment of students' work is criterion referenced and is designed to be both formative and summative. In practical terms, this requires teachers to assign skills descriptors that measure a student's progress in each of three dimensions: knowledge, experience, and interpersonal. Support for primary teachers who faced the challenge of implementing the reform was limited to various documents and exemplars provided by CDC. From early meetings with teachers in focus groups, those of us working in TELEC discovered that there was some confusion about how teachers might go about implementing the new curriculum at classroom level. We decided that we would design the databases in such a way that they clearly reflected the thinking behind TOC yet could stand alone if the TOC were ever dispensed with for any reason.

In the early stages of database preparation, we were aware that one of our biggest challenges would be to encourage teachers to be more student centered and to adopt a communicative approach. Bearing this in mind, our objective was to provide materials that do not promote passive learning but rather that will encourage students to learn experientially and will appeal to teachers as "doable." These materials are complemented by a range of professional development resources.

TeleNex Primary comprises two hypertext databases, *PrimeGram* (a pedagogic grammar) and *PrimeTeach* (graded task-based teaching materials and information on ELT theory). Figure 11.4 shows some of the files available in *PrimeTeach* using the *Topics* entry point.

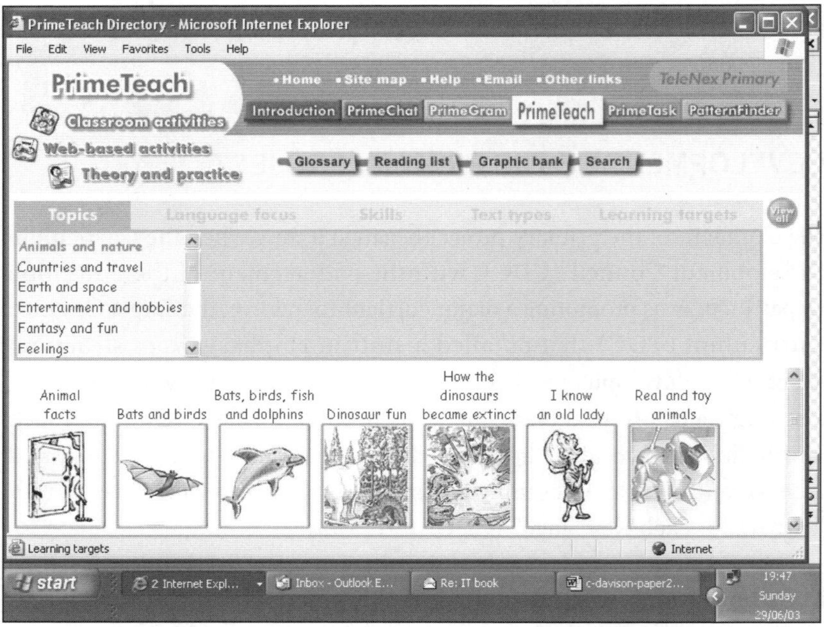

Figure 11.4 Directory screen in *PrimeTeach*

To supplement the teaching materials, interactive multimedia tasks (*PrimeTask*) are available in CD-ROM format, designed for use in the computer laboratory under the direction of the English teacher. *PrimeTeach* and *PrimeTask* aim to help teachers with the implementation of task-based learning and the integration of information technology into the English curriculum, at Key Stage 2 (KS 2), i.e., Primary 4–6.

Drawing on five years of experience in website and ALN development, the TELEC team were able to construct the primary databases on the basic model of the secondary databases. However, we were not content merely to replicate the secondary website. Rather, we wanted to build on the demonstrated strengths of the established databases and exploit the potential revealed by the ever-changing medium. In designing *TeleNex* Primary, we have taken into account the reality of the Hong Kong primary teaching context together with the greater flexibility of information technology. Our main concern has been to make the resources accessible, appealing, useful, and user-friendly. To ensure quality control, TELEC staff have conducted periodic meetings and workshops with a focus group of practicing primary teachers, as well as visiting a wide range of primary classrooms over the three-year development period.

CONSIDERATIONS IN DEVELOPING A PEDAGOGIC REFERENCE GRAMMAR

PrimeGram aims to provide English teachers at primary level with information about English grammar that will enhance their understanding of and proficiency in English and help them teach more effectively. The design criteria had therefore to take into consideration the knowledge background of the target users, and our initial research indicated that teachers are inclined to see grammar as a set of seemingly arbitrary rules rather than as a set of resources for meaning making. The challenge is to help teachers develop a more coherent framework for thinking about grammar than a traditional/structural approach that emphasizes the forms and rules of grammar. To this end, we have written the grammar files from a functional perspective, drawing on the work of Halliday (1985).

This approach takes as its starting point the assumption that every time we use language, we do three things simultaneously: we talk or write about some*thing*, i.e., we use language to talk about our experience of the world around us; we talk or write to some*one*, i.e., we use language to interact with other people; and when talking or writing, we make *links* between the elements of language in order to create whole units of meaning, or coherent texts. With reference to the language system, there are a number of resources that cluster around each of these three dimensions, which Halliday refers to as *metafunctions*: the *experiential* metafunction (resources for talking about the world), the *interpersonal* metafunction (resources for interacting with others), and the *textual* metafunction (resources for creating cohesive and coherent texts). From a pedagogical perspective, viewing language in relation to these three dimensions provides a useful way for teachers to introduce elements of the language system to students.

Adopting a functional approach involves a number of changes in emphasis, from seeing language as a set of rules, to seeing language as a set of resources for making meanings; from seeing language as something we know, to seeing language as something we do; from a preoccupation with syntax, to a focus on semantics; from sentence level to whole texts. Despite these changes in emphasis, users who are familiar with a more traditional formal approach to grammar will find *PrimeGram* accessible. Any terms that are likely to be new or difficult to understand are glossed with clear explanations and examples. For further information on the pedagogical implications of a functional approach, see, for example, Celce-Murcia (1991); Butt, Fahey, Feez, Spinks, and Yallop (1994); Eggins (1994); Hasan and Perrett (1994); Lock (1996); and Derewianka (1998).

Another key consideration is that most primary teachers are non-native speakers of English, Cantonese being their mother tongue. An awareness of the problems with English that many teachers have as a result of first language transfer has been taken into account at all stages of development. Terminology has proved to be a particularly "knotty" issue, given that (a) our target readers are not native speakers of English, and (b) the adoption of a functional model brings with it a number of unfamiliar specialist terms. Our approach has been to write at all times in a clear, accessible style; and, when the use of technical terms is necessary, a gloss icon is used to trigger a popup containing a definition plus example and links to other key terms. A related terminological issue involves providing guidance for teachers when introducing metalanguage in the classroom. We encourage teachers to adopt a step-by-step approach. For example, in introducing *adjectives*, the first step is to use everyday descriptions and explanations: "This is a word that describes something." The next step might be to use an everyday label: "This is a word that describes, so we can call it a *describing word* or a *describer*." For certain grammatical areas, students will need to know the technical term that is used in their textbooks. When students understand the meaning and function of the grammatical item, it is suggested that teachers can introduce the technical term: "We have been using the term *describing word*, but now we're going to learn the proper term: *adjective*."

There are many reference grammars on the market, and it may be asked why yet another is necessary. The problem with most grammars is simply that they cater for a very general readership, and as such they often cannot answer the specific questions that local English teachers have. With these points in mind, *PrimeGram* seeks to provide a comprehensive pedagogic reference grammar in which accurate grammatical information is presented in a simple, clear, and lively way, relevant to the Hong Kong primary ELT context and customized for a target user whose mother tongue is Cantonese. Although we have aimed for comprehensive coverage, we recognize that not all of the information provided is relevant for teaching in the primary years one to three, and four to six (KS1 and KS2). Therefore, we draw explicit attention to those areas that are most relevant to teaching the language at this level.

To summarize, in the early stages of development, we faced a number of special challenges. As a result of having addressed these, *PrimeGram* is unique in the following ways: It caters especially for English teachers at primary level, it has been developed specifically for the Hong Kong context, it uses both a traditional and functional approach, it exploits the latest multimedia technology and is available on the Internet, and it is interactive.

This last item points to one of the key differences between *PrimeGram* and a standard reference grammar. Teachers are encouraged to engage in experiential activities, to comment on and query what they have read, and to discuss aspects of language structure and usage in the Conference Corners. It is felt that through engaging with the grammar interactively in this way, teachers are likely to develop a better understanding than through passive reading.

PRIMEGRAM: ARCHITECTURE AND CONTENT

PrimeGram comprises three main sub-components: *communicative functions*, *text types*, and *grammar*.

These sub-components appear as entry points on the opening screen for *PrimeGram* and suggest to the user three different perspectives on English: a functional focus, a focus on whole texts, and a structural focus. For the first-time user, a tutorial provides the theoretical rationale for adopting this approach, and *Help* files provide guidance in navigating through the different options and levels.

The *Grammar* sub-component constitutes the core of *PrimeGram*. Preliminary research indicated that teachers have little understanding of clause structure and its relevance to English language teaching. This was our first challenge, and our response was to use the directory as an aide-mémoire for the rank scale: sentence/clause/groups and phrases/word class. This approach makes clear that sentences are made up of clauses, which in turn are made up of groups and phrases, which in turn are made up of individual words. By clicking on an option in the directory, a user accesses an overview file for that level of the grammar, and a further list of options opens up on the sidebar menu. For example, at clause level, the three main options are:

- *Using the clause to represent experience* (experiential metafunction);
- *Using clauses to interact* (interpersonal metafunction);
- *Using the clause to organise text* (textual metafunction).

Each choice opens up a further range of options; for example, if a user clicks on *Using the clause to represent experience*, the options now available are:

- *Different types of experience*
- *The participants: 'Who or what is involved?'*
- *The circumstances:'How?', 'When?', 'Where?'*
- *The processes 'What's happening?'*

Through discussion with the focus group, we found that users are initially more comfortable dealing with grammar at word class level. The directory thus caters for both novice and sophisticated users. Novice users can enter at the familiar word class level and are then encouraged to follow links from here to other parts of the grammar as appropriate. Once engaged in the files at word class level, it is hoped that the user will follow links to files that deal with groups and phrases and so on, up the hierarchy. It is envisaged that after a period of working within the *Grammar* component, teachers will begin to use previously unfamiliar entry points and, in so doing, gradually build up an internal picture of clause constituency.

We have endeavoured to make the grammar descriptions clear, interesting, relevant, and appealing. Our approach has been to write in a way that places a minimal processing load on the user, through avoidance of nominalizations, etc., and by ensuring that grammatical explanations are profusely illustrated with simple examples from appropriate sources. To this end, we have created a cast of characters (the Gong family and friends), fictional but recognizably based in Hong Kong. These characters appear in three narrative texts, excerpts of which are used for illustrative purposes, together with graphics where appropriate. Feedback from our focus group indicates that the grammar pages are appealing and psychologically reassuring for the user, who can expect to encounter familiar faces and scenarios while exploring unfamiliar linguistic territory. In addition, animation, video, sound files, and graphics are exploited wherever they can help to put across difficult concepts effectively. For example, when introducing suprasegmentals in the context of a discussion on mood, audio is used with animation to illustrate intonation contours and stress patterns.

While the *Grammar* sub-component can be seen as the core of *PrimeGram*, the *Communicative functions* sub-component is designed as an alternative entry point into the grammar. Users can choose from three options:

- *Talking about experience*
- *Interacting with others*
- *Creating links in text.*

The first set of files, *Talking about experience* (which maps on to Halliday's experiential metafunction), is concerned with the language choices available when we are talking or writing about the world around us, e.g., talking about people, places, and things, what is happening, who is doing what, and the circumstances surrounding these activities (when, where, why, how). The second set, *Interacting with others* (the interpersonal metafunction), is concerned with the language choices we have when we communicate with others in our daily life. This interaction can be through speech or writing, and involves formulaic expressions for such things as greeting people. It also involves the range of expressions for requesting and providing information, asking people to do things for us, and offering to do things for others, expressing feelings and opinions, and adjusting our interaction with others, e.g., through modality. Information is also provided about topic management at text level, i.e., starting and maintaining a topic, changing the topic slightly, introducing a new topic, etc. The third set, *Creating links in text* (the textual metafunction), contains information about the ways in which text is made cohesive, and emphasizes the difference between spoken and written language. Topics covered include reference, ellipsis, substitution, conjunction, the intricacies of lexical cohesion, and thematization. The three areas within *Communicative functions* provide a straightforward functional overview of the areas suggested by each metafunction. As indicated earlier, viewing language in relation to these three dimensions provides a useful way for teachers to introduce elements of the language to students. And if students can also be encouraged to think of there being three main functions of language, then they will find it easier to use English language to do things in real life. The files within the *Communicative functions* sub-component are written in such a way to encourage the user to explore other areas of *PrimeGram*. For example, after reading about different ways of organizing information in texts (dealing with thematization), the user has a number of options, including following a link to where thematization is dealt with at clause level in the grammar; or, the user can choose to explore how clauses are tied together to create particular text types such as descriptions or recounts.

The *Text types* sub-component is designed to deepen teachers' understanding of how different text types are structured and how grammar functions in the context of complete texts (in contrast to a sentence-level approach). It is envisaged that this understanding will encourage teachers to work with a wider range of texts and explore new approaches to presenting grammar in the classroom. Within the *Text types* component, there are ten main areas:

- *Descriptions (General and Particular)*
- *Recounts*
- *Instructions*
- *Stories*
- *Reviews*
- *Explanations*
- *Arguments*
- *Texts for social interaction*
- *Texts for fun and leisure*
- *Text formats for presenting information*

The first seven areas illustrate the social purpose, schematic structure, and typical grammatical features of each general text type, using sample texts that are appropriate for the primary classroom. By clicking on one of these options in the sidebar menu, the user accesses a second level that contains a list of the files available dealing with that particular text type. Each set contains four core files: an overview of the structural and grammatical features, a file containing typical examples (which are recycled throughout the set and exploited for analyses), a file that analyzes the schematic structure, and a file that outlines the key teaching implications. In addition to these core files, each set contains a number of files that deal with the characteristic grammatical features associated with each text type. Our aim is to show how each text can be seen as a configuration of meanings, and the reader's attention is constantly drawn to how the overall shape of a text emerges from the way meanings cluster at certain stages as the text unfolds. Links are made to related sections of the *Grammar* sub-component for teachers to follow if they want to know more about a particular aspect of grammar. Teaching suggestions are also provided with links to *PrimeTeach* where appropriate.

The last three areas in the *Text types* directory are collections of more specific forms of texts referred to in the KS2 syllabus. These are conversational texts (phone calls, service encounters, etc.); texts for enjoyment (poems, plays, jokes, etc.); and text formats (tables, charts, diagrams, etc.).

As labels for text types, the items in the *PrimeGram* directory are somewhat problematic. The problem can be expressed in simple terms: what constitutes a genre or text type? There is little agreement in the literature, and Fairclough (1996, 13) for example, notes that much of the confusion surrounding the terminology can be traced to "a common failure to distinguish different levels of abstraction." He suggests that a primary distinction can be made between "actual texts and the conventions which people draw upon in producing and interpreting them." He goes on to propose a secondary distinction within these conventions, between "text types and the more abstract constituents of text types" such as genre and discourse. The work done by genre analysts such as Swales (1986), Dudley-Evans (1989), and Bhatia (1993) suggests that we recognize genres as salient entities because of their prototypical schematic structure. However, such work does not go much beyond stage and move analysis. Pioneering work done in Australia by Martin and Rothery (1986), Hammond (1987), and Christie (1989) has focused more on lexico-grammatical features of different text types. This approach encourages the analyst to view each text type (e.g., a narrative, a particular description) as having a distinctive shape or structure. The shape or structure of a narrative or particular description then is shared by a set of texts in which meanings are configured in the same or similar ways related to their respective social purpose, be it of narration or description. This work has had a significant impact on English language teaching; see, for example Derewianka (1990) and Feez and Joyce (1998). The experience of introducing texts at primary level in Australia has informed the development of the *Text types* sub-component, but inevitably, we have had to compromise between adhering to the terminology deriving from current research and our awareness of the existing knowledge and expectations of Hong Kong primary teachers. The directory of text types that are available to the user in the sidebar menu have been chosen with three major objectives in mind. Firstly, they should represent the types of writing which students are required to comprehend and produce, as indicated in the TOC guidelines. Secondly, they should refer unambiguously to recognizable text types. Thirdly, and perhaps most importantly, they should encourage teachers to move beyond the rather narrow confines of the sorts of texts traditionally produced in the English classroom at primary level, i.e., very short texts which seldom move beyond narration and description.

NAVIGATION WITHIN *PRIMEGRAM*

Before considering the navigational problems associated with a hypertext grammar, it may be worth pausing to examine briefly the nature of hypertext itself — in particular, how it differs from traditional linear text. Fiderio (1988) explains that hypertext, at its most basic level, allows the user to connect screens of information using associative links. Schneiderman and Kearsley (1989) define a *hypertext* as a database that has active cross-references allowing the reader to "jump" to other parts of the database as desired. Other definitions reflect the different preoccupations of the various disciplines involved in hypertext development: education, literature, computer science, and cognitive science [see Balasubramanian (1997) for a diverse range of definitions]. The practice of jumping from one bit of text to another is taken for granted by website users; however, the manner in which the links are established and the way the connections are perceived by users are key issues in hypertext design. Hypertext gives users the flexibility to read files in any order, according to their own agenda. This imposes a tremendous constraint on the materials writers, because, unlike a book that is designed to be read linearly, each page of a hypertext document has to be written with the assumption that a potential reader has not necessarily read what precedes it. Balasubramanian (1997) notes that in a complex hypertext database, users must be able to move freely through the system according to their needs, without getting lost spatially or cognitively. In an attempt to preempt "hyperspace vertigo" in users, we have given a great deal of thought to the mechanisms by which users will find their way within files and across the sub-components. Reduced to basic figures, *PrimeGram* contains some 3059 html files, distributed across three main components, each comprising multiple areas and sub-areas. An idea of the complexity of *PrimeGram* can be seen in Figure 11.5.

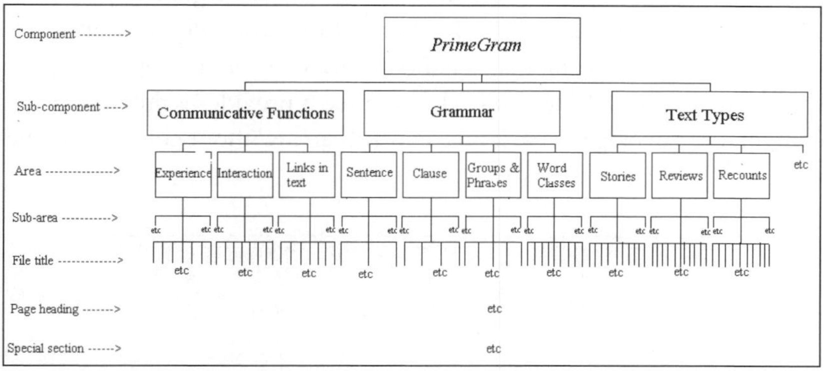

Figure 11.5 *PrimeGram* sub-components

For the user, the problems of navigation can be narrowed down to the following questions: Where am I now? Where have I come from? (And how do I return?) Where can I go next? What is the big picture?

In order for the reader to be able to answer the above questions at any point within the grammar database, we have paid due attention to the nine considerations for writing hypertext promulgated by Schneiderman and Kearsley (1989): chunking (i.e., organizing the information into small chunks that deal with one topic, theme, or idea, each chunk representing a node in the hyperdocument); interrelationships; consistency of document names; master reference list; simplicity in traversal; screen design; low cognitive load; early reviews; and maintenance of multiple perspectives.

What, then, are the mechanisms that we have put in place to ensure users' spatial and cognitive orientation?

For the big picture, a site map provides a schematic representation of the relationship between each component. In orientation, the look and feel of the different components are designed to assure the users at all times that they are still in the familiar *TeleNex* environment. Color is used systematically to identify each major component (e.g., yellow for *PrimeTeach* and blue for *PrimeGram*). At a macro-level within *TeleNex* Primary, users can navigate easily from one component to another by clicking on the macro-navigation buttons in the permanent tool bar at the top of the screen (see Figure 11.6).

When first accessing *PrimeGram*, the user arrives at an opening screen and has the option of accessing any one of three main sub-components through the following options: Text types, Communicative functions, or Grammar.

Once a sub-component has been selected, a tree diagram appears that allows the user to preview the further options available for that level, without leaving the opening screen. Rolling over any of these options opens up a further menu of file titles. This approach provides a preview of the content and coverage of each area and sub-area without committing the user.

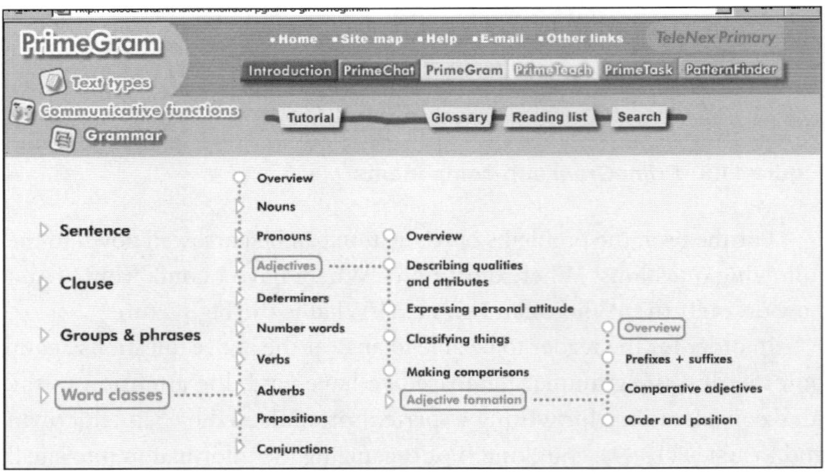

Figure 11.6 ***PrimeGram*** **directory screen: Grammar**

If the user rolls over *Groups and phrases*, then a further menu of options appears: *Noun group, Verb group, Adverb group, Prepositional phrases,* and *Adjective group*. Alternatively, the user has the option at any point of clicking on an item and going to a content page at any level in the hierarchy. On arriving at the content page, the user sees the three main sub-components with the current sub-component highlighted, and to the left side of the content page is a sidebar menu with the options listed and the current file highlighted (see Figure 11.7).

The sidebar menu provides a list of the available areas and sub-areas within each sub-component. Each item on the sidebar menu leads to a file, which may have a number of pages and sections. The sidebar menu fulfils an important navigational and orienting function. If the current

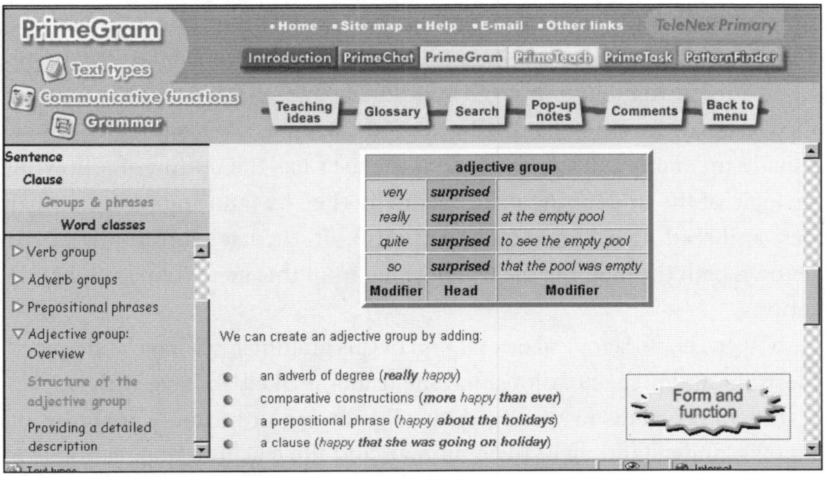

Figure 11.7 ***PrimeGram*** **Content Page with Sidebar Menu**

component is the *Grammar*, the user can see at a glance where he or she is in rank scale (sentence/clause/groups and phrases/word class), and the current file is indicated through highlighting. The user can also see the options available at that point. Similarly, the sub-component indicator fulfils a dual navigational and orienting function. If the user wishes to confirm which is the current component or visit one of the other components within *PrimeGram* (*Communicative functions, Grammar, Text types*), he or she can refer to the top left-hand corner of the screen.

Within *PrimeGram*, the information has been distributed in such a way to avert information overload for the user. With respect to chunking (Schneiderman and Kearsley 1989), each file has been broken down into pages (i.e., scrollable screens) which are effectively the small chunks dealing with a particular topic, theme, or idea. The objective is to provide short scrolling screens; however, in some special cases when the writer considers it preferable not to fragment the information, longer scrolling screens are used. Where it is deemed in the reader's interest to be offered a jump to a related file, a standard rubric is used, followed by the file title and pathway, indicated through the standard convention of underlining. Because of the complexity of the grammar, the information is layered in such a way that the user does not feel threatened or overwhelmed. Clicking on a file title in the sidebar menu leads to a *front page* in which the key information is collected in digest form. This information is typically a summary of the main grammatical points, with an indication of its relevance to the rest of the grammar and to teaching. This front page includes graphics and appropriate illustrative examples plus an invitation for the user to pursue the area in greater depth through additional pages that are listed in a menu of page-heading options at the bottom of each page. Within a file (which is made up of any number of pages), the user has the option of following the logic of the underlying print hierarchy (i.e., by ignoring links to other files, and reading each page sequentially); or, the user can follow his or her own path through the file by choosing from this menu of page-heading options.

Given the pedagogical orientation of this grammar, we have anticipated questions and preoccupations that the reader is likely to have with respect to student problems in specific areas of grammar, uncertainty about his or her own understanding of the grammar, and the teaching implications for presenting particular items and areas in the classroom. At various points throughout a file, the user has the option of clicking on a range of buttons labelled: *Student problems, Try it out!, Teaching suggestions, Form and function, Misconceptions,* and *Analysis.*

Clicking on one of these buttons takes the user to a special section, each of which is outlined as follows:

1. The *Student problems* sections focus on typical productive and receptive problems. Each section contains an outline of the problem plus illustrative examples, with an explanation about why students are likely to make such a mistake. Reference is made where appropriate to a corpus of learner writing. Analyses reveal interesting and useful patterns of usage: overuse, underuse (or avoidance), errors (lexical, collocational, and syntactic), and of course correct usage.
2. The interactive *Try it out!* activities are provided to engage teachers' interest, to consolidate teachers' knowledge, and provide for some discreet practice in a non-threatening environment. They are not intended for use with students, but it is suggested that teachers might like to develop a similar activity (not necessarily multimedia) using it as a model.
3. The *Teaching suggestions* sections provide an outline of teaching implications plus examples of exercises, suggested contexts, activity ideas, board work, animations to illustrate intonation patterns, etc., plus links to appropriate units of work in *PrimeTeach.*
4. The *Form and function* section is used throughout to emphasize that there is never a one-to-one relationship between form and function: One grammatical form might perform a number of different functions,

and a particular function might be expressed by a range of grammatical forms, e.g., an adverb group can be variously used to provide information about circumstances (e.g., how, when, where, why), or to express modality and personal viewpoint, or to link sentences as conjunctives/connectives. Alternatively, where the focus is on function/meaning, the *Form and function* section can be used to show the various forms that can realize that function/meaning, e.g., how circumstance can be realized by adverbs, adverb groups, prepositional phrases, and adverbial clauses.

5. The *Misconceptions* section is used to outline misconceptions about English, particularly those that are being brought to light by corpus linguistics studies: see for example, Sinclair (1991) and Biber, Johansson, Leech, Conrad, and Finegan (1999). Coverage in a *Misconceptions* section may also be given to teachers' misunderstanding of a particular area of grammar, as revealed, for example, through the messages posted in the language conference corner.
6. The *Analysis* section provides an interactive text analysis whereby the reader can see particular features of a text highlighted by using the mouse to roll over key words. This approach has been found to be particularly effective in *Text types* files: for example, when dealing with the text structure of a particular description, rolling over the stage names highlights the section of text under focus and pops up a commentary. Likewise, when dealing with the different grammatical features of each text type, this approach allows the user to see each instantiation of a given feature throughout a whole text (plus annotation in the form of labelling, arrows, etc.), simply by rolling over a key word such as *comparison* or *contrast* (see Figure 11.8).

Spreading the information across the files through this device of clearly labelled buttons reduces the cognitive load on the user and ensures that users know exactly what they are getting when they click on a button. Information that is considered to be of secondary importance is available in popup windows via *note*, *glossary*, and *example* icons. At all times, we have endeavoured to write files from the end user's perspective. In order to decrease the burden on the user's short-term memory, all screens have a consistent layout and have been designed so that the reader has an immediate perceptual grasp of the current file type and of the current sub-component. This is of particular importance in orienting the sort of user who will take full advantage of the reading flexibility which hypertext offers.

Here is the general description of bats. Please roll over to see the language used to **compare** and contrast.

	Bats
IDENTIFICATION	Bats are mammals. They are flying mammals. They are the only mammals which can fly. There are two types of bats, mega bats and micro bats.
DESCRIPTION **appearance**	Bats have brown or black fur. Bats have small pointed faces and large ears. They have strong chests and shoulders but they have thin hips and legs. They also have large wings. The wings of bats are made of skin, or membrane, which stretches over long finger bones. On the end of the thumb there is a claw.
type: mega bats	The wings of mega bats are bigger than their bodies. Mega bats have good eyesight. Fruit bats and flying foxes are mega bats. These bats live in the tropical forests of India, Africa, Asia and Australia. Flying foxes are the largest bats in the world.
type: micro bats	The wings of micro bats are smaller than the wings of mega bats. Micro bats have good hearing. They make short high-pitched sounds which bounce off nearby objects. Then they listen for the echo in order to 'see' where they are going. This is called echolocation. There are micro bats all over the world. The smallest micro bats in the world live in Thailand. They are called bumblebee bats.
diet	Most bats are insectivorous. They eat large quantities of insects so they help farmers and gardeners. Fruit bats and flying foxes eat fruit and flowers. In South America there are vampire bats which feed on the blood of mammals and birds.
behaviour	Bats are nocturnal. They fly and search for food at night. During the day they sleep, or roost. When they roost, bats hang upside down by their feet. They usually roost inside dark, cold spaces, for example, inside caves or old buildings. Some bats, however, roost outside on trees or rocks. Usually bats fly and roost in groups. A group of bats is called a colony. Bats are very clean. Every day they clean their teeth and wings, and they lick their fur like a cat.
life cycle	Female bats usually bear one infant at a time. The infants have no fur. They are also blind and deaf. They grow quickly and become adults in two months. Some bats can live for thirty years.
human opinions	In the West people often think that bats are scary and mysterious because they are nocturnal animals, while in Asia people think that bats mean good luck and long life.

Figure 11.8 Interactive reading popup (rollover effect on "contrast")

In the course of development, *PrimeGram* has been extensively trialed with focus groups of practising and pre-service teachers whose feedback has allowed us to fine-tune the database by content and navigational ease. This means that, in spite of the complexity inherent in a database of this type, a user is able to identify the precise location conceptually within the grammar (at the level of word, group, clause, sentence, or text) and, at a practical level, within the website itself.

THE QUESTION OF SUSTAINABILITY

Corbel (this volume) points to sustainability as a key issue in an enterprise of this nature, arguing that only a small number of education projects that receive funding result in continuing services after funding has ended. He argues that websites in particular are treated as discrete products, whereas their need to be maintained and continuously developed makes them more like services. The problem is that the maintenance and development of such a service requires continued funding for technical, clerical, and database support. Lack of recognition of this need by educational systems is a major concern that could undermine the investment of time and money into IT support and professional development systems such as *TeleNex*.

Since its inception in 1993, *TeleNex* has developed from a dial-up proprietary network with only 300 users to a website which has over 10,000 users, the majority of whom are teachers in secondary and primary schools in Hong Kong.

Funding for the development of *TeleNex* expired at the end of 2001, which has meant that TELEC is no longer able to maintain technical or database staff. However, under the aegis of HKU's Faculty of Education, the *TeleNex* website remains active on the Internet. Registered teachers are able to access all of the databases and Faculty of Education staff join the discussions from time to time on a voluntary basis. Newcomers to the Hong Kong English language teaching community are welcomed as *TeleNex* users.

REFERENCES

Allan, Q. (1998). Enhancing the language awareness of TeleNex teachers through corpus data. *Journal of Technology and Teacher Education, 6*(4), 57–74.

Allan, Q. (2003). The TELEC Secondary Learner Corpus: A resource for teacher development. In S. Granger, J. Hung, and S. Petch-Tyson (eds.) *Computer learner corpora, second language acquisition and foreign language teaching* (pp. 195–211). Amsterdam: Benjamins.

Balasubramanian, V. (1997). *State of the art review on hypermedia issues and applications.* Graduate School of Management, Rutgers University, Newark, NJ. Retrieved July 13, 2003, from http://www.win.tue.nl/2L670/static/chapter0.html.

Berry, R. (1994). Using concordance printouts for language awareness training. In C. Li, D. Mahoney, and J. Richards (eds.) *Exploring second language teacher development* (pp. 195–208). Hong Kong: City University Press.

Bhatia, V. (1993). *Analysing genre: Language use in professional settings.* London: Longman.

Biber, D., Johansson, S., Leech, G., Conrad, S., and Finegan, E. (1999). *Grammar of spoken and written English.* London: Longman.

Bourne, J. (1998). Net-learning: Strategies for on-campus and off-campus network-enabled learning. *Journal of Asynchronous Learning Networks, 2*(2), 70–89.

Butt, D., Fahey, R., Feez, S., Spinks, S., and Yallop, C. (1994). *Using functional grammar: An explorer's guide.* Sydney: National Centre for English Language Teaching and Research.

Celce-Murcia, M. (1991). Grammar pedagogy in second and foreign language teaching. *TESOL Quarterly 23*(3), 459–79.

Christie, F. (ed). (1989). *Writing in schools: Study guide*. Geelong: Deakin University Press.

Curriculum Development Council (1997). *Syllabuses for primary schools: English language primary 1–6*. Hong Kong: Hong Kong Government.

Derewianka, B. (1990). *Exploring how texts work*. Sydney: Primary English Teaching Association.

Derewianka, B. (1998). *A grammar companion for primary teachers*. Sydney: Primary English Teaching Association.

Dudley-Evans, T. (1987). *Genre analysis: An investigation of the introduction and discussion sections of MSc dissertations*. In M. Coulthard (ed.) *Talking about text*. University of Birmingham: ELR Discourse Monograph No. 13.

Eggins, S. (1994). *An introduction to systemic functional linguistics*. London: Pinter Publishers.

Fairclough, N. (1996). *Critical discourse analysis*. London: Longman.

Feez, S. and Joyce, H. (1998). *Writing skills: Narrative and non-fiction text types*. Sydney: Phoenix Education.

Fiderio, J. (1988). *A grand vision. Byte, 13*(10), 237–44.

Halliday M. (1985). *An introduction to functional grammar*. London: Edward Arnold.

Hammond, J. (1987). An overview of the genre-based approach to the teaching of writing in Australia. *Australian Review of Applied Linguistics, 10*(2), 163–81

Hasan, R. and Perrett, G. (1994). *Learning to function with the other tongue: A systemic functional perspective on second language teaching*. In T. Odlin (ed.) *Perspectives on pedagogical grammar.* Cambridge: Cambridge University Press

Johns, T. (1988). Whence and whither classroom concordancing? In T. Bongaerts, T. de Haan, S. Lobbe, and H. Wekker (eds.) *Computer applications in language learning* (pp. 1–16). Dordrecht: Foris.

Lock, G. (1996). *Functional English grammar: An introduction for second language teachers*. Cambridge: Cambridge University Press.

Lock, G. and Tsui, A. (1996 July). *Grammar and the teacher: Building an electronic grammar database for secondary ESL teachers*. Paper presented at the 23rd International Systemic Functional Congress, Sydney.

Martin, J. and Rothery, J. (1986). *Writing project report* (Working Papers in Linguistics 4). Department of Linguistics, University of Sydney.

Schneiderman, B. and Kearsley, G. (1989). *Hypertext hands-on!: An introduction to a new way of organizing and accessing information*. Reading, MA: Addison Wesley.

Sengupta, S. and Nicholson, S. (1996). Online and ongoing: Teacher development through TeleTeach. *English Language Teaching Journal, 50*(4), 290–302.

Sinclair, J. (1991). *Corpus, concordance, collocation.* Oxford: Oxford University Press.

Swales, J. (1986). *Genre analysis: English in academic and research settings.* Cambridge: Cambridge University Press.

Tsui, A. and Bunton, D. (2000). *The discourse and attitudes of English language teachers in Hong Kong. World Englishes, 19*(3), 287–303.

Tsui, A. and Ki, W. (1996). An analysis of conference interactions on TeleNex: A computer network for ESL teachers. *Education Technology Research and Development, 44*(4), 23–44.

Tyrwhitt-Drake, H. (1999). Responding to grammar questions on the Internet: Providing correction through the corpus. *English Language Teaching Journal, 53*(4), 281–8.

Wu, K. and Sengupta, S. (1998). Hong Kong ESL teachers' questions about English grammar. *Language and Education, 12*(2), 137–53.

Wu, K. and Tsui, A. (1997). Teachers' grammar on the electronic highway: Design criteria for TeleGram. *System, 25*(2), 169–83.

12

The Standards Solution: The Role of Standards in Developing IT Professional Competencies and Communities of Practice in English Language Teaching

Chris Davison

> **standard**, n. 1. Distinctive flag; rallying principle. 2. Weight or measure to which others conform or by which the accuracy of which the others is judged. 3. Degree of excellence required for particular purpose. 4. Thing recognised as model for imitation. (*The Concise Oxford Dictionary* 1999)

INTRODUCTION

A key question in the literature on IT and educational innovation is how to achieve change in teacher practices. It seems relatively easy to achieve superficial or one-off changes in teacher behaviour, but longer-term sustainable and significant change seems much more difficult. This is true in many areas of education, but in IT and education, the mismatch between expectations and results, between investment and return, seems overwhelming (see, for example, Cuban 2001). Given government demands for accountability and increasing use of standards in education as a way of achieving school-based reform, this chapter explores the extent to which the application of IT teacher standards might contribute to the development and evaluation of IT professional practice in English Language Teaching (ELT).

The notion of a "standard" is somewhat ambiguous, as revealed by any recourse to the dictionary. One interpretation is that of "setting a standard," establishing a rallying principle, exemplified by the calls to become IT-literate, to meet the demands of the information age. Another interpretation is that implied in the phrase "raising standards," indicating the degree of excellence required for a particular purpose. The term "standard" also suggests a model for imitation, as in "following standards." A final interpretation is measuring or judging the extent to which others conform, as in "meeting the standard." These different, even conflicting, interpretations of what is meant by standard are particularly problematic in new or emerging fields of professional practice, such as IT and ELT, in which expectations for the field are still being articulated and negotiated.

This chapter first critically examines the origins of the demands to establish IT teacher standards generally and then looks at the background and development of the competency-based movement, both in the ELT and in the IT fields. A number of different IT competencies or standards frameworks are then examined, including examples from the US, the UK, Australia, and Hong Kong. Adopting a view of language and learning as socially constructed (Halliday 1993; Vygotsky 1978) and a notion of community of practice derived from the work of Lave and Wenger (1991) and Hung and Chen (2001), the potential benefits and limitations of the establishment of IT standards in ELT for developing a community of professional practice, and for enhancing and evaluating that practice, are then explored. The chapter concludes with suggestions for approaches to professional development and evaluation that incorporate more recent post-modern perspectives on teacher desire and on complexity and innovation in educational organizations (Fullan 2003; Hargreaves 1994).

RESPONDING TO THE IT IMPERATIVE

As Benson (this volume) notes, educational systems and governments around the world are investing heavily in IT infrastructure as well as professional development, whole schools and educational systems now networked and online. The urgency to respond to this so-called IT imperative is demonstrated in government statements such as that by the Australian National Office for the Information Economy (1998, 11):

> Our education and training systems must equip all Australians to be enterprising, innovative, adaptable and socially responsible participants in the information economy. There is an urgent need to recognize that Australia will be at a serious disadvantage in the global knowledge

> economy if it fails to produce workers, professionals and managers with the skills to work in the online environment.

The pessimistic rhetoric of the National Board of Employment, Education and Training (NBEET) (1998, xiii) goes further, raising concerns that Australia may have not adopted IT processes and thinking quickly enough, nor at a high enough level, claiming, "Australia can not simply afford to be a spectator" but may already "be condemned to be an also-ran compared with other nations." They recommend that all educational systems "should examine the quality of IT education at the primary and secondary levels with a view to implementing the world best practice in terms of IT and teacher competence" (NBEET 1998, xiii). Australian governments are not alone in appearing convinced that more, and better, IT use in schools is essential for further economic development, for employment, and hence, for quality of life.

Given the enormous costs to educational systems of maintaining and upgrading IT hardware, it is not surprising that teachers are continually exhorted to embrace IT and to "raise their standards" of knowledge and skill as well as application, as in an article from an Australian national newspaper, entitled "Study warns teachers on IT abilities" (Lawnham 2002, 11):

> Teachers and trainee teachers should be formally required to reach minimum standards of computer competency ... there is clearly strong national support from this from all Australian education systems.

The report of this study, entitled *Raising the Standards,* proposes a national framework for ICT competence, building on a number of state-based initiatives in the area, to be elaborated later in this chapter. Similar dire warnings of a coming Dark Age if teacher IT competence is not urgently addressed proliferate the media of most developed, and even many developing, nations.

There are two main reasons for this global phenomenon: firstly, the widespread view that unless teachers themselves experience and foster a technological workplace that replicates those in the government, business and community sectors, they will be unable to teach the values, beliefs, attitudes and skills their students will need to successfully cope with the information age (Australian Council for Computers in Education 2003). Hence, it is assumed that IT is a cross-curricular issue, that all teachers need to be not only IT-literate but actively using IT in their daily classroom practice. The second reason for the concern with teacher standards is the assumption that IT will, almost single-handedly, revolutionize the state of teaching, and hence, by definition, the state of the "knowledge" society (Castells 1998; Brown and Duguid 2000). This rather erroneous and naïve belief is so pervasive in the media and in government that it needs further examination.

Although international research has shown that IT acts more as lever than as a trigger or catalyst for pedagogical, and societal reform (OECD 2002), unrealistic expectations among policy-makers and the wider community about the impact of IT are well documented (see Benson, Cummins, this volume). The grassroots assumption is that IT has the potential to irrevocably alter the patterns of classroom interactions and student learning. However, as the Australian Council for Computers in Education (2003) points out:

> Political structures that say that investment in technology-supported classrooms in isolation will alter learning are misguided. Improved learning is a product of teachers using valid and appropriate strategies that are formed by an understanding of the available educational resources including technology. Teachers cannot do this from a technology-bereft position and without professional development to help them build a learning technology process.

The same point is made by Taylor (1994, 184), who argues:

> It is entirely feasible to surround a teacher with a team of audio-visual technicians, graphic artists and computing specialists to vary the style of the delivery of the educational message without producing a significant increase in pedagogical efficacy. The key process for improving the quality of teaching and learning is instructional design.

In other words, using ICT as a mode of delivery will not automatically improve student learning. IT in itself cannot transform ELT classrooms, but embedded within a constructivist pedagogy (see, for example, Cummins, this volume), it can be used by the informed and confident teacher to boost English language learning outcomes significantly. For example, IT use in ELT enables the building of powerful interactive networks and connections to information and people around the world (see, for example, Legutke, this volume; Cummins and Sayers 1995; Kourtis-Kazoullis and Skourtou, in press) which can not only greatly enhance communicative purpose and student motivation but also significantly increase opportunities to use English in a range of authentic and meaningful contexts. IT also provides much greater flexibility to build English language knowledge and skills than the traditional teacher-centered textbook-dominated classroom can offer, as demonstrated by Love (this volume). If we look at IT as a social resource to create meaning, as in a systemic-functional linguistic model of language learning, we can see the potential of the IT medium in new ways. For example, using the ever-increasing resources of the Internet as well as digitalized information, IT can be used to build knowledge of the field — the subject matter of the English language texts and their related concepts and vocabulary — in a much more elaborate and individualized way. Using asynchronous communication networks, including email and web forums, IT can also expand, even transform, the tenor of relationships between teachers and students, and between students themselves, through offering a enormous variety of opportunities to adjust the type, frequency and quantity of the interaction and interactants. Finally, despite its yet inadequate inability to handle real-time spoken conversation (see Markee, this volume), IT offers significantly increased opportunities to exploit the full mode continuum, providing an enormous diversity of text-types, including those in other languages, and greatly increased communicative opportunities ranging from most colloquial and context-embedded situations to the most formal de-contextualized situations.

Given the potential of IT to add value to learning and teaching, it is perhaps understandable that both investment and expectations in new learning technologies are high and rising. If, in response to their investment, governments and educational systems see little or no uptake by teachers and no overt change in pedagogical practice, there is a tendency to blame the teacher. This leads to the widespread belief in the community, even among teachers themselves, that many classroom practitioners are resistant to technological change, that they are opposing, even subverting the IT imperative, in order to cling to their old out-dated pedagogies. Lists of IT teacher competencies and standards frameworks are seen as a justifiable bureaucratic response to this apparent lack of teacher knowledge, skill and even motivation, especially in situations where school-based reform is seen as urgent and IT skills as essential for economic advancement. However, although standards can be a useful tool in developing and evaluating IT in ELT, they are not without their problems. The following section of this chapter briefly describes the competency-based movement and the development of standards in the IT and in the ELT fields, before exploring their potential benefits, and inherent limitations, for establishing, developing and evaluating IT professional practice in English language education.

THE COMPETENCY-BASED MOVEMENT: THE STANDARDS SOLUTION

The competency-based movement is a global phenomenon, fuelled by the rise of economic rationalism and corporate managerialism in government. Concerned about moribund or failing economies, from the 1980s onwards, many governments around the world enthusiastically adopted the tools of private enterprise to create competition in public organizations to achieve micro-economic reform. They developed outcomes-based measuring tools to establish and compare gains in achievement. The measuring tools are supposed to give "clients" or users clear information about effectiveness, hence the capacity to make an informed choice about where to go to "buy" what they need. Competency-based frameworks are seen as essential for governments to set achievement standards and to know who is achieving or not achieving according to these standards. They also give government the opportunity to tie funding to indicators of achievement.

The incorporation of competency-based frameworks or standards into education has become an entrenched feature of the educational systems of the English-speaking world. According to the International Board of Standards for Training, Performance and Instruction (IBSTPI), a competency involves a related set of knowledge, skills and attitudes that enable a person to effectively perform the activities of a given occupation or function in such a way that meets or exceeds the standards expected in a particular profession or work setting (Richey et al. 2001).[1] Typically, a competency is divided into specific indicators describing the requisite knowledge, skills, attitudes and context of performance. In the US, for example, legislation requires all states to establish standards or indicators of program quality for primary, secondary and adult education curricula, and to explicitly link performance standards or criteria to these indicators. More recently, the concept of mandatory standards has been extended to the development and credentialing of teachers as well as programs.

The ELT field has also been a fertile ground for the development of such teacher standards, perhaps because of its diversity and its relatively recent development as an area of professional activity. Such teacher standards describe the knowledge, skills and attitudes that teachers should be able to demonstrate within certain defined areas. According to Nunan (in press):

> This performance-based approach to teaching and teacher education stands in contrast with the traditional *credentialing* approach in which the certificates, diplomas, and degrees that a teacher holds form the criterion for determining an individual's professionalism.

In Australia, a set of teacher standards or competencies were developed in the early 1990s to provide advice to employers on "the development and implementation of induction and on-going professional development" (Strong and Hogan 1994) as well as to guide teacher training institutions in the development of pre-service and in-service education courses in ELT. In Europe, the University of Cambridge Local Examinations Syndicate (UCLES) has developed performance standards in a number of areas for teachers at different stages of professional development (for example, see University of Cambridge Local Examinations Syndicate 1996). In North America, the international TESOL professional association has developed a framework and set of teaching standards for teachers of adult ESL students. As can be seen in Table 12.1, this framework establishes performance standards and sample performance indicators for teachers in eight domains of teaching: planning, instructing, assessing, identity and setting, language, learning, content, and professionalism (TESOL 2002).

Table 12.1 TESOL standards for teachers of adult learners (TESOL 2002, adapted from Nunan, in press)

Domain	Standard	Sample performance indicators
1. Planning	Teachers plan instruction to promote learning and meet learner goals, and modify and adjust instruction plans in relation to learner engagement and achievement.	• Identify and articulate learning goals for both language and other content. • Design short-term and long-term plans to promote learning. • Select appropriate resources.

2. Instructing	Teachers create supportive environments that engage all learners in purposeful learning and that promote respectful interactions among learners and between learners and their teachers.	• Create physical and virtual environments that engage all learners. • Organize and manage constructive interactions among learners. • Engage learners in decision-making about their learning.
3. Assessing	Teachers recognize the importance of and are able to gather and interpret information about learning and performance to promote the continuous intellectual and linguistic development of each learner. Teachers use knowledge of student performance to make decisions about planning and instruction "on-the-spot" and for the future.	• Gather, interpret, and document information about learner performance before, during, and at the end of instruction. • Engage learners in self-assessment. • Use assessments that allow learners to demonstrate their learning. • Use assessment instruments that are equitable.
4. Identity and setting	Teachers understand the importance of who learners are and how their communities, heritages, and goals shape learning and expectations of learning. Teachers recognize the importance of the socio-cultural and socio-political settings — home, community, workplace, and school — that contribute to the identity formation and therefore influence learning. Teachers use this knowledge of identity and settings in planning, instructing, and assessing.	• Create an environment conducive to adult learning. • Establish classroom routines that encourage learners' appreciation for each other. • Model impartial attitudes towards cross-cultural differences and/or conflicts. • Take information from learners' communities to guide planning, instructing and assessing.
5. Language	Teachers demonstrate proficiency in English	• Demonstrate proficiency in oral, written, and nonverbal English. • Serve as English language models for learners.

6. Learning	Teachers draw on their knowledge of language and adult learning to understand the processes by which learners acquire a new language in and out of classroom settings. They use this knowledge to support adult language learning.	• Create situations in which meaningful messages are exchanged. • Provide learning experiences that promote autonomy and choice. • Provide learning experiences that respond to differential rates of learning.
7. Content	Teachers use their understanding of the structure and function of language to support language learning. Teachers also use their understanding of the connections among concepts, procedures, and applications from content areas relevant to learners to further learners' language development.	• Design contextualized activities to provide practice with English discourses and texts. • Provide input and practice of socially and culturally appropriate language. • Use content as a vehicle for language instruction.
8. Commitment to professionalism	Teachers continue to grow in their understanding of the relationship of second language teaching to the community of English language teaching professionals, the broader teaching community, and communities at large, and use these understandings to inform and change themselves and these communities.	• Take information from the communities at large and the broader teaching community to inform the English language teaching profession. • Develop personal professional development plans. • Reflect on teaching practice to continue to grow professionally.

Interestingly, there is little or no mention of IT competencies in this comprehensive conceptualization of an ELT teacher's work, with the exception of a single reference to a teacher's ability to "create physical *and virtual* environments that engage all learners" (Standard 2: Instructing, my emphasis), although work has recently commenced to develop a set of IT standards for the TESOL profession. The same notable absence of IT knowledge and skills applies to the teacher standards in the Australian and European examples described earlier. Such an omission could be indicative of a relative lack of interest in the development of IT skills in the ELT field, which appears unlikely given the increasing number of textbooks and research reports on the topic. Alternatively, it could be symptomatic of a deeper problem with the application of IT to ELT, suggesting, albeit implicitly, practical limitations to the new medium in teaching English language, both as a pedagogical tool and a medium of communication. Or, paradoxically given the focus of this chapter, it could simply be a reflection of the ELT field's priorities; that is, there are far more important knowledge and skills for an English language teacher than those related to IT. The reality is probably a mix of these three factors, which suggests that we need a closer look at what might be involved in being "competent" in IT.

As alluded to above, many educational systems have developed and/or adopted their own standards-based frameworks for IT competence, which are being used not only for professional development but also for credentialing purposes. For example, in the US there has been significant emphasis on modifying teacher pre-service programs to ensure that new teachers are prepared to use ICT in the classroom. Recent national initiatives include the US Department of Education's *Preparing Tomorrow's Teachers* program, and the *Teacher Preparation StaR Chart: A Self-assessment Tool for Colleges of Education*, released by the CEO Forum on Education and Technology. Several sets of standards have also been developed by the National Council for Accreditation of Teacher Education, to be used in evaluating teacher preparation programs for accreditation, the most significant being *Recommended Foundations in Technology for All Teachers* and *Standards for Basic Endorsement in Educational Computing and Technology Literacy* (NCATE/ISTE 2000).[2]

At the state level there is even more activity. Georgia, Kentucky, Texas and Virginia have all developed plans to address the technological education of new teachers. The *A Plus Education Reform Act of 2000*, approved by the Georgia legislature in July 2000, requires trainee teachers to be proficient in computer and other instructional technology applications and skills. In Kentucky, accredited teacher training institutions are required, to ensure student teachers meet the new standards, including those for technology, set by the Kentucky Professional Standards Board. The State Board for Educator Certification in Texas has established standards for all new teachers, regardless of teaching field or subject certification. In Virginia, the State Board of Education requires that all teachers will have to demonstrate proficiency in technology for license renewal, starting in 2003. Other states have also begun implementing technology standards for teachers.

In the UK, all trainee teachers are required to meet mandatory standards in ICT, set by the Teacher Training Agency, in order to achieve qualified teacher status, and by 2002, it was assumed that "serving teachers should generally feel confident, and be competent to teach, using ICT within the curriculum" (The Independent ICT in Schools Commission 1997). As indicated earlier, in Australia a national framework for ICT competence has been proposed which would generate specific task standards and make explicit the knowledge, attitudes and skills that make up competency (Department of Education 2002). However, there has already been extensive work on the development of IT standards in the various Australian states. For example, in Queensland, minimum standards form part of an enterprise bargaining agreement for teachers. Professional development funding has been directed to assisting teachers to achieve the mandated standards. The framework also includes non-mandatory descriptors. In Victoria, all trainee teachers are required to meet the standards established in the *Learning Technologies Capability Guide* (Victoria) (2000). This standards framework identifies and summarizes six areas for skill development in the use of learning technologies for classroom and administrative purposes. In Hong Kong, mandatory ICT standards for trainee teachers have also recently been introduced.

Despite all this activity, no systematic attempts to develop specific IT standards for English language teachers, or indeed for any other specific subject area other than IT itself have occurred.[3] Some possible reasons for this state of affairs in ELT were alluded to above, including lack of knowledge and/or interest, and other competing priorities. Those involved in the development of IT competencies suggest a justification for this: research shows that the skills needed by teachers are the same across subjects and levels of schooling; hence, only generic standards, albeit with subject-specific examples are needed (Lawnham 2002; Australian Council for Computers in Education 2003). A closer look at some of the benefits and limitations of IT teacher standards in ELT, however, shows that this view may need to be partially qualified.

The next section of this chapter looks at the ways the application of IT teacher standards could be helpful in the ELT but also some of the problems that might be encountered.

IT STANDARDS IN ELT: POSSIBILITIES AND PROBLEMS

The application of IT standards to ELT has a number of potential benefits, including the establishment of a shared set of expectations or practices for

IT in ELT (i.e., *setting* standards); the articulation of a clear set of stages for the development of teacher IT competence, which might be used as a guide for professional development programs or for independent learning (i.e., *raising* standards); and the explicit recognition of achievement or progress in the development of IT competencies among the individual teachers or even organizations (i.e., *meeting* standards). These potential benefits, and the problems they raise, are addressed in the following section. A particular issue that is foregrounded in the discussion is the inherent contradictions implied in *following* standards in such a rapidly changing IT environment, in which teachers are being exhorted to transform their pedagogy and develop their own professional development goals to take them into an uncharted information age.

Setting standards

One of the most obvious benefits of IT standards in ELT is that it should facilitate the development of a shared set of expectations or practices for IT in ELT, and a common language for talking about those practices. Such standards would theoretically make it explicit to teachers what knowledge, skill and attitudes are involved in becoming pedagogically IT-literate, thus enhancing the development of a community of IT practice. Communities of practice, as defined by Lave and Wenger (1991), consist of informal social organizations resulting from mutual human engagement in joint enterprises. Over time, participants in these communities come to share ways of doing things, forms of interaction and discourse, beliefs and values, and ensembles of resources that include tools, concepts, and symbols.

In contrast to approaches that conceive of learning as an isolable experience in which individuals engage in order to gain specific knowledge and skills, Lave and Wenger (1991, 51) emphasize "learning, thinking, and knowing are relations among people in activity in, with, and arising from the socially and culturally structured world." Teacher practice, inside and outside the classroom, is shaped both by the external context, that is, the institutional, socio-cultural and political practices of the society in which the teacher operates, and by the complex interactions between learners and between teachers and learners. Thus, in any construct of professional knowledge, theoretical and pedagogic knowledge interact with and are influenced by other sorts of implicit or explicit contextual knowledge, including institutional, socio-cultural and political knowledge. Together, these epistemologies or knowledges could be perceived to form teachers' professional knowledge and are both reflected in and shaped by practice.

Membership of the community of professional practice is gained by a lengthy process of socialization in which individuals acquire not only the knowledge and skills of the group — its particular professional knowledge and expertise — but also its distinctive cultural ethos, values and attitudes. The procedures, rules and criteria by which professional judgments are made, however, tend to remain implicit and obscure to non-members. In this sense, professional knowledge or expertise appears esoteric in character.

Many researchers claim that the explicit delineation and communication of the underlying theoretical concepts, models and frameworks which inform and guide practice is essential for the credibility, viability and quality of professional practice, particularly in areas such as ICT where technology and delivery mechanisms are evolving rapidly. As Garrison (2000, 3) argues in relation to distance education:

> While those who are experts in the practice of a variety of forms of distance education may have the tacit knowledge to intuitively guide their educational decisions and effectively facilitate learning, this is not sufficient for the vast majority in the field.

Competency-based frameworks seem to fulfill the required function, that of making tacit pedagogical knowledge explicit. However, there are a number of problems with this simple equation. They include concerns about the definition, nature and scope of IT teacher competencies, and the interaction between competence and context.

The first problem, that of defining IT teacher competencies, includes not only the issue of what competencies to include but also the nature and scope of those competencies. It is highly likely that English language teachers will consider pedagogical-related IT competencies as far more important than those competencies considered peripheral to their classroom practice. For example, Scheffler and Logan (1999) found that general computer competencies were rated relatively low in importance by teachers, the highest ratings in that domain being given to knowledge of how computers assisted individuals, the ability to demonstrate keyboard skills, use of proper terminology to communicate about computers, proper care and use of computer systems, use of computers in decision making and use of computer reference materials. In contrast, knowledge and skills in how to make computers an integral part of the school and classroom environment were rated much higher. The majority (twenty-six of the thirty-four) of the competencies ranked as important or very important were those classified as "computer skills unique to teachers." While this may not necessarily reflect what teachers actually need, it does indicate that pedagogical issues are foremost in teachers' minds when they reflect on the use of ICT. It is for this reason that some educational systems have reconceptualized IT in education as *learning technologies*. The argument is that the term better captured the holistic and process-oriented nature of IT use in learning and teaching, the reality that students are engaged in educational activities while using a range of technological tools and processes. The term encompasses both old technologies such as pen and paper, and new and potential technologies such as outlined by Murray (this volume).

This reconceptualization of IT raises the question of whether there are, in fact, any differences between IT teaching competencies and general teaching competencies in ELT. As indicated above, ELT teacher competencies and IT teacher competencies have generally been considered separately, as has been the case with most specific subject areas. Cummins (this volume) has highlighted the lack of integration of information technology with instructional approaches in teaching content standards, leading to teachers teaching to the test, or regressing to outdated behaviorist methodologies in IT, such low-level drill and practice type activities. ELT as a still-developing area of professional practice is even more vulnerable to such tendencies.

More integration between ELT and subject content is certainly long overdue, but many IT competencies are also unique to the medium. For example, there are certain technical skills and requirements for successfully moderating and facilitating online discussions and chat sessions (e.g., Collison et al. 2000; Kearsley 2000; Rosenburg 2001). In online asynchronous discussions, teacher-moderator competencies include ensuring adequate time for learner reflection, eliciting a range of contributions and keeping contributors on task, and organizing and storing discussions to be used in subsequent lessons. In online synchronous discussions (e.g., chat), the moderator competencies include establishing ground rules for discussion, prompting interactions with minimal teacher intervention, sensing how online text messages may appear to less engaged learners, and responding to cultural differences. In addition, teachers need to encourage a critical perspective in their students, an understanding of the socially and politically constructed nature of these new technologies, and "the impact of global connectivity on individuals, nation-states and global well-being" (ACCE 2003). Many of these pedagogic demands will seem familiar and achievable to ELT teachers, but the ways in which a teacher demonstrates such competencies will be quite different from the traditional ELT classroom in which time and space are held constant. The vast range of IT delivery environments (classroom-based, Internet-based, laboratory-based based, hybrid environments, and so on) also makes the pedagogic demands more complex and demanding. Even the most constructivist and culturally sensitive ELT classroom is a different environment with such IT use, hence the need for additional but integrated competencies for ELT teachers adapted to their new roles and circumstances.

This raises the second major problem in setting IT teacher standards, that of formulating the statement of competency in such as way that captures the complexity of the competence required. Reynolds and Salter (1995) list three models of competence in teacher training. In the first model, competence is represented as a list or combination of discrete tasks. Tasks are analyzed into components, and each component part is stated as desired behavior. A competent teacher is one who can perform the behaviors involved in the pre-specified tasks. This model appears to be exemplified by standards which are presented as long decontextualized checklists. Such a narrow view of competency defies common sense and professional practice. The second model of competence focuses on the ability to transfer previous learning to new situations, and is more dynamic and interactive. The third and most desirable model approaches competence as the application of a combination of knowledge, understanding, experience and executive ability to perform a range of tasks in specific contexts. Such a model does not represent competencies as a list of technical skills, or even pedagogical and curriculum skills, but emphasizes the connection among the content, roles and circumstances of teaching.

The problem, as ACCE (2002) argues, is that competencies are generally associated with highly formalized professional practices and are not usually applied to ill-defined tasks involving variable and uncertain circumstances, procedures and outcomes. Ill-defined tasks certainly include many forms of IT and English language teaching. In response to this kind of argument, some educational systems such as Victoria's have adopted the term "capabilities" to emphasize that competencies are not simply a testing device but involve professional judgment and evaluation. A closer examination of their competency framework reveals some key characteristics of their indicators of competence, including an emphasis on open-ended process and collaboration, an ability to learn through "through exploration and play," and a holistic view of the use of IT in the classroom (see Table 12.2).

Table 12.2 IT teacher standards in Victoria

Area 6 — Using learning technologies in the key learning areas
Stage One **Identify and explore a range of learning technologies for teaching and learning in the key learning areas** Visit a leading practice classroom to examine how learning technologies are integrated into key learning area programs; Research SOFWeb and the Internet for literature related to using specific learning technologies in the key learning areas, for example, Leading Practice IdeaBank; Become familiar with a specific learning technology through exploration and play; Undertake a professional development activity related to a specific application, for example, through subject association or Leading Practice Classrooms Network activities.
Stage Two **Incorporate the use of appropriate learning technologies in key learning area programs** Devise a personal project that will develop skills in the use of a specific learning technology; Work with a team to develop strategies for integrating use of learning technologies in key learning area programs; Trial a Leading Practice IdeaBank unit of work and provide feedback; Develop a unit of work that integrates the use of a specific learning technology in a key learning area program; Participate in a Global Classroom project specific to a key learning area.
Stage Three **Investigate opportunities for innovative use of learning technologies in key learning area programs** Undertake an action research project with a group of other staff to extend the possibilities for use of a specific learning technology; Publish an account of your experience in the use of a specific learning technology in a professional journal; Conduct a conference workshop on the use of a specific learning technology; Share experiences of using a specific learning technology in the classroom with a teacher network; Undertake leader training for Computing across the Primary Curriculum (CAPC) or Computing across the Secondary Curriculum (CASC); Contribute to Leading Practice IdeaBank.

(Extracted from *Victorian Learning Technologies Capabilities Guide,* 2000)

A third problem with setting standards centers on the role of context in the interpretation of competencies. A common assumption underlying competency-based approaches is that standards can be teacher- and context-free, but many studies show that standards are interpreted differently by individuals according to their personal background, previous experience, unconscious expectations, internalized and personalized preferences regarding the relative importance of different standards and ideological orientation. As Gipps (1994, 167) suggests, the central problem is that "we are social beings who construe the world according to our values and perceptions." The generalized and relative nature of much standards-referenced assessment schemes makes conflicting interpretations even more likely. Research suggests, however, that standards always need interpretation (Brindley 2001) and statements are always ambiguous and require the application of implicit knowledge (Claxton 1995; Wiliam 2001). In other words, standards may provide a basis for defining and negotiating pedagogic activities, but ultimately teachers will only understand and interpret standards by looking at actual examples of classrooms and activities. Even if it were possible to define standards unambiguously, it is not necessarily desirable. As Wiliam (2001, 171), a vocal critic of standards-referenced systems argues: "Put crudely, the more precisely we specify what we want, the more likely we are to get it, but the less likely it is to mean anything." The assumption that teachers will accept externally imposed standards as a basis for action is also problematic, given the increasing evidence that when a conflict arises between standards and teachers' own personalized judgments, teachers manipulate and/or reject the indicators of performance (Davison 1999, 2004).

The greatest problem with setting IT teacher standards in ELT, however, is the dearth of meaningful empirical research to guide and enhance practice that could provide a sound and comprehensive basis to set standards for professional practice, or judge the quality of such standards. As Cummins (this volume) points out, "despite the increase in access to IT in schools, there is still no consensus on *how* computers and other new technologies should be used to support learning." Hence, the articulation of a clear set of stages for the development of teacher IT competence, to be used as a guide for professional development programs or for independent learning (i.e., raising standards), is somewhat problematic. This issue is elaborated in the next section.

RAISING STANDARDS

The second major set of benefits usually associated with the development of IT teacher standards is the provision of a clear set of stages for the development of teacher IT competencies, which can then be used as a guide for professional development programs or for independent learning. Research clearly shows that teachers' capacities to make effective use of IT in educational settings are undermined by inadequate knowledge and skills, as well as by a shortage of properly trained instructional designers and educational support personnel. The development of IT teacher competencies in ELT ideally should result in much improved training, as desired outcomes and stages in the development of those outcomes are made more explicit, even though the continual and rapidly evolving nature of IT, and the associated escalation of system-level initiatives and expectations, will make the development of such competencies a continuous process. There are two related problems, however, with relying on competencies as tool for professional development: firstly, the mismatch between competencies and the nature of teachers' professional development processes; and secondly, the problem of support.

In relation to teachers' professional development processes, research suggests that teachers develop professional knowledge from practice up, rather than education down, through "personally constructed meanings" (Yaxley 1991, 87) derived from their unique experiences, individual conceptions, unexamined values and beliefs, and interactions with local contexts, which are shaped by and, in turn, shape classroom instruction. Effective professional development is assumed to necessitate a recognition and reappraisal of values, beliefs and attitudes as well as teaching practices, and an attempt to reconcile inconsistencies and develop greater coherence and applicability (Burns 1992; Connors 1991; Walker and Evers 1984; Yaxley 1991). As is elaborated in Davison (2001), Eraut (1986) distinguishes four paradigms of professional development: defect, growth, change and problem solving. These models appear to parallel Zeichner's (1983, 6) classification of beliefs about learning to teach: the behaviorist paradigm, the growth paradigm, the craft or apprenticeship approach and the inquiry oriented paradigm. The models assume particular views of the nature of professional competence or knowledge and the function of professional development, and each is associated with particular professional development strategies (and models of curriculum and program development).

The first model of competency outlined by Reynolds and Salter (1995) above — the checklist of discrete skills — seems to fit within Zeichner's (1983) behaviorist model of learning to teach, professional development in Eraut's terms, aimed at redressing defects to overcome teacher obsolescence or inefficiency. Such professional development seems to be essentially product oriented, professional knowledge being externally constructed and delivered to the teacher through checklists and professional development "packages" with little, if any, negotiation in content selection with teachers. As Jones and Moore (1993, 388) argue, such a narrow view of competencies transforms the mode of control over teacher expertise from cultural to technical:

> By decontextualizing 'skills' and abstracting them from their constitutive cultural practices, these reductive procedures construct partial, disembedded representations of the complex social interactions (involved in teaching) ... a set of operations to be implemented and followed as instructed. Its practitioners have no automatic access to its theoretical base or means of critical examination.

The deficit model fails to recognize teacher competence as a dynamic construct; hence, it may foster professional development practices that are static and unresponsive to different contexts and changing teacher needs. Emphasizing top-down delivery can also breed over-dependence on outside agencies as the sources and mediators of change.

Reynolds and Salter's (1995) second model of competency seems to fit within Zeichner's (1983) craft or apprenticeship model of learning to teach. Professional development, in Eraut's (1986) terms, is aimed at change. The change model sees the system and/or the profession as constructing the professional knowledge rather than the individual teacher. According to Eraut (1986, 733) "such changes will not necessarily be recognized, understood, or desired by all teachers but they will have to come to terms with them." The assumption is that the teacher will fit the change, rather than vice versa. The model is linked with program evaluation in which teacher performance and development are seen as part of an overarching agenda. Teacher development and school development are intertwined (Fullan 1991, 2003). This model is associated with professional development delivery through the introduction of curriculum and standards frameworks by the system. The teacher is "apprenticed" to the structures rather than to a master-teacher, and through models and exemplars is "developed." Connors (1991, 58) points out that this model assumes three general outcomes of effective professional development for the teacher; changes in teachers' beliefs and attitudes, changes in teachers' instructional practices and changes in teachers' content knowledge.

However, he also highlights the importance of understanding the teacher's perspective on change. The fact that the nature, timing and rate of change is driven by the system rather than by the teacher, means that as a model of professional development it can get too far ahead of teachers and lose its purpose and relevance for them. Hargreaves (1994, 11) highlights a central problem with both the defect and the change paradigms:

> Political and administrative devices for bringing about educational change usually ignore, misunderstand or override teachers' own desires for change. Such devices commonly rely on principles of compulsion, constraint and contrivance to get teachers to change. They presume that educational standards are low and young people are failing or dropping out because the practice of many teachers is deficient or misdirected. The reason why many teachers are like this, it is argued, is that they are either unskilled, unknowledgeable, unprincipled or a combination of all three.

A more effective match between the desired model of competence and the view of learning to teach and professional development is clearly needed. Eraut's (1986) third model of professional development is growth, which views the practitioner as a self-motivated and reflective teacher seeking fulfilment through self-improvement. However, this model sees professional knowledge as almost entirely experiential and internally developed, emphasizing self-evaluation and teacher-selected content. It is associated with unstructured teacher discussion groups, collaborative workplace professional development and individual reading. Such professional development may suit innovators or early adopters in IT (Geoghegan 1994) but seems unlikely to be the preferred paradigm of the majority. The other problem with the growth model is that it can lead to a lack of uniformity in IT standards across a profession and result in wide discrepancies in classroom practice.

Eraut's (1986) fourth model of professional development is the problem-solving paradigm, associated with inquiry-oriented beliefs about learning to teach. This model assumes that the act and the context of teaching are complex and problematic but that solutions are best worked out by teachers in their own classrooms and workplaces. As Connors (1991, 56) emphasizes, teachers are in the best position to identify and solve problems collaboratively, with external consultants or mentors' help as needed. Professional development in this paradigm is achieved through active involvement in the problem-solving process. The model constructs professional knowledge as experientially based, as in the growth model, but there is much more explicitness about the nature of that knowledge and the processes of its development and a recognition of the value-saturated nature of teaching. Critical reflection is the central strategy associated with the problem-solving paradigm, although as Calderhead (1991, 3) notes, there are a number of problems with an over-reliance on reflection, including difficulty in describing development, the problematic relationship between personal knowledge and curricular knowledge with the inevitable privileging of the latter, the inherent contradiction between the need to develop particular areas of knowledge, skill and attitudes and the aim of encouraging autonomy and professional responsibility within a professional set of obligations, the need to reconcile a focus on reflection with demands for teacher assessment, and the problems of individual differences in beliefs and learning practices. The dilemma can be summarized as how to ensure teachers acquire certain knowledge and skills while still allowing them to construct their own knowledge.

A fifth conceptualization of professional development perhaps more suited to the development of the more complex desired model of competency is one which sees professional knowledge as being jointly constructed and both experiential and theory-driven. Professional development and change takes place through being confronted with and having to articulate different ways of doing and seeing the same thing, through conflict and critique, not just collaboration (or co-option), friction not just reflection. In the process, teachers are forced to articulate their own professional understandings, to critique others, and, in so doing, are better prepared for both the pedagogic and practical demands of working with IT. This approach enables grassroots' needs, experiences and reactions to feed into ongoing IT policy development more systematically and gives teachers a greater sense of control over educational change. The competencies themselves should also stimulate self-reflexivity, requiring teachers to examine the relationships between pedagogy and technology, to ask key questions about the nature of the pedagogical options available for use of IT, and the extent to which IT can enhance the impact of different approaches to pedagogy. Most importantly, the framing of the competencies should be *futures-oriented* to ensure teachers develop a pattern of lifelong learning in relation to IT.[4]

The second major problem with using competencies to raise IT standards is the need to provide adequate teacher support. Teacher stress, lack of collegiality and leadership, and the scale and pace of change all can undermine the effectiveness of IT professional development. Stressed teachers do not have the energy to take risks, to try out new ideas, to accept the inevitable, if temporary incompetence, that accompanies experimentation. They adopt self-protection strategies rather then open themselves up to change. Research on teacher development shows that teachers must be acknowledged as competent professionals (Logan and Sachs 1988) and given both support in the personal risk taking that professional development demands and a sense of control over the process (Fullan 1991, 2003). The nature of ELT is such that teachers are often divided and isolated, even from others who face the same pedagogic challenges. Hence, teachers also need opportunities to observe colleagues, exchange feedback, consult and reflect with colleagues and plan and evaluate together (Anastos and Arcowitz 1987; Eisner 1988; Little 1984). In addition, teachers need access to mentor figures and leaders in the field who they can turn to for guidance and support (Ingvarson and MacKenzie 1988). Teachers also need access to ideas about how workplaces, community groups, government groups, business and industry use information technology. They need support to identify the processes involved that are akin to processes they teach or should teach, irrespective of the age range of their students.

The final factor inhibiting professional development and change is the most critical one of all — time. Research shows again and again that developing new knowledge and skills and integrating new skills into practice takes a lot of time, three to five years at least. Collins (1991, 19) argues, "it takes a lot of courage and support through months of feeling clumsy ... a long practice time" before a fundamental breakthrough point is reached when teachers understand why this works better (Fullan and Hargreaves 1992) and integrate the approach or idea into their own personal theory of teaching (Little 1984). The rate and scale of change in IT, exacerbated by pressure to meet political, rather than professional, timelines has been a particular and obvious stumbling block to successful implementation of IT in ELT at all levels. In particular, many ELT teachers may be worried if they "will have the opportunity to learn enough to meet standards" (Lawnham 2002, 11).

Meeting standards

The third main potential benefit of IT teacher standards is to provide explicit recognition of achievement or progress in the development of IT competencies among individual teachers or even organizations (i.e., meeting standards). It also can provide Geoghegan's (1994, 11–4) compelling reason for teachers to adopt the innovation, albeit through a "big stick" approach to professionalism (ACCE 2002), as well as make institutional commitment explicit. In the competency-based movement, there are different ways to validate that a person has demonstrated the relevant competencies; but in addition to the problems already discussed above, all suffer from privileging one view of pedagogy and teacher competence over others, and from the problem of inference.

For the first problem, even a cursory examination of the standards and performance indicators presented in Tables 12.1 and 12.2 reveal their ideological biases. As Nunan (in press) notes in relation to the TESOL standards, indicators such as *engage learners in self-assessment, model impartial attitudes towards cross-cultural differences and/or conflicts, provide learning experiences that promote autonomy and choice, develop personal professional development plans*, selected more or less at random, reveal that individual responsibility and self-direction, as well as tolerance for pluralism are important values for the creators of these standards, and it is against these that TESOL teachers are to be judged. The notion of privileging raises more challenging questions about the learning resources considered appropriate for participation in IT-based classroom practices, such as whether the teacher is defined as the sole source of knowledge in IT practice or if peer input is also considered a valued learning resource, whether "standard English" is defined as the only appropriate code for IT practice or whether ESL learners' evolving linguistic repertoires, including interlanguage forms, are also viable resources for IT participation and meaning making. This links to the wider issue of the non-neutrality of technology: what do ICTs privilege, what do they disallow and how does using them influence what is included and excluded in the curriculum (see Lankshear and Snyder 2000, for a fuller discussion of ICT non-neutrality).

The need to meet predetermined standards also raises the controversial issue of ownership and control. As Jones and Moore (1993, 391) argue, competency-based frameworks regulate access and control through the regulation of outputs (performance indicators) rather than through inputs (knowledge and cultural attributes acquired through professional socialization). This is a particular concern in such a rapidly evolving field as ICT, which demands that teachers look beyond Murray's (this volume) old concepts and metaphors in order to create new pedagogic possibilities.

Self-reflexivity and a futures-orientation, incorporating the ability to plan for new unimagined possibilities, have to be built into any set of IT competencies in ELT. As Hawkins and Irujo (1994, 13) point out in relation to ELT, the paradox is that through competencies:

> We are trying to create a sense of community in our (teacher) learners. A stable community, however, inhibits change. If we are successful in building a strong community, do we then lose the ability to change when necessary?

The second problem of whether one can infer underlying competence from samples of individual observable behavior has troubled the field of language assessment for decades. The difficulty is compounded by the fact that IT teaching in particular can be seen as so interactive and collaborative that assessment of individual teacher competency seems impossible. Hence, in adopting or developing any IT teacher standards in ELT, standards should neither be used as conditions of employment nor strategies to measure teachers' eligibility for professional advancement. Teachers unable or unwilling to be evaluated should not be disadvantaged or considered unprofessional on this issue alone. As ACCE (2003) argues:

> A good teacher does not stop being a good teacher simply because they have not demonstrated competency in the use of learning technology. However, good teachers in the 21st century can reasonably be expected to be, amongst other things, good users of learning technology. The development of confidence in adapting to new ideas and technology is as important as the development of the competencies themselves.

CONCLUSION

Most ELT teachers value their professional skills very highly and are motivated to do all they can to provide their students with every pedagogic opportunity to succeed. Such teachers usually feel almost a moral obligation to incorporate IT into their pedagogic practice when faced with evidence that IT can enhance linguistic and learning outcomes, especially in the context of high employer and professional expectations for IT use. However, many feel that their relatively limited exposure to IT has left them poorly prepared to handle perhaps even basic processes such as establishing web forums. Even if they master these skills, the pedagogical challenges that are involved in integrating IT in the ELT curriculum are deep-seated. For many teachers with little or no access to the necessary technology, technical support and professional development structures, the expectation that they will achieve a high level of IT competency can be intimidating, even undermining. For many such teachers, the existence of "competencies" creates a benchmark against which they know they will be evaluated poorly and perhaps even be labeled "incompetent." As a result, any attempt to define or mandate competencies will only gain acceptance within the ELT profession if it is approached in an atmosphere of trust and respect, and supported by provision of the necessary time, resources, equipment, support and professional development to allow all teachers to gain the necessary competencies to add to their pedagogic repertoire.

NOTES

1. For all practical purposes, the terms *competencies* and *standards* are now used interchangeably in the literature to refer to the same phenomenon; hence, I have not attempted to distinguish between them.
2. Developed jointly by the National Council for Accreditation of Teacher Education (NCATE), the official body for accrediting teacher preparation programs in the US, and the International Society for Technology in Education (ISTE), the professional education organization responsible for recommending guidelines for accreditation to NCATE for programs in educational computing and technology teacher preparation.
3. In most secondary schools, especially at the senior levels, ICT is also offered as a subject area in its own right with its curriculum and assessment standards.
4. Teachers' needs vary over time and they may require different kinds of support and resources at different stages of their professional development. In Fielding's (1983) terms, as the teacher begins to move towards becoming a novice professional, more apprenticeship-like approaches may be needed, but as the novice professional moves into the student-centered stage and starts to be more analytical and critical, more reflective approaches may be appropriate.

REFERENCES

Anastos, J. and Arcowitz, R. (1987). A teacher-directed peer coaching project. *Educational Leadership, 45*(3), 23–42.

Australian Council for Computers in Education. (2003). Teacher Learning Technology Competencies Discussion Paper Revision 1.3. Retrieved July 14, 2003, from http://www.acce.edu.au/tltc/b-contents.asp.

Australian National Office for the Information Economy. (1998). *A strategic framework for the information economy: Identifying priorities for action.* Canberra: Commonwealth of Australia.

Brindley, G. (2001). Outcomes-based assessment in practice: Some examples and emerging insights. *Language Testing, 18*(4), 393–407.

Brown, J. and Duguid, P. (2000). *The social life of information.* Cambridge, MA: Harvard Business School.

Burns, A. (1992). Teacher beliefs and their influence on classroom practice. *Prospect, 7*(3), 56–65.

Calderhead, J. (1991). *Reflective teaching and professional development.* Schools Program Division, Ministry of Education and Training, Victoria.

Castells, M. (1998). *End of millennium.* London: Blackwell.

Claxton, G. (1995). What kind of learning does self-assessment drive? Developing a 'nose' for quality; comments on Klenowski. *Assessment in Education, 2*(3), 339–43.

Collins, C. (1991). Teacher development: Achievements and challenges. In P. Hughes (ed.) *Teachers' professional development* (pp. 10–24). Melbourne: Australian Council of Educational Research.

Collison, G., Elbaum, B., Haavind, S., and Tinker, R. (2000). *Facilitating online learning: Effective strategies for moderators.* Madison, WI: Atwood Publishing. ED 448 684

Connors, R. D. (1991). Teacher development and the teacher. In P. Hughes (ed.) *Teachers' Professional Development* (pp. 53–81). Melbourne: Australian Council of Educational Research.

Cuban, L. (2001). *Oversold and underused: Computers in the classroom.* Cambridge, MA: Harvard University Press.

Cummins, J. and Sayers, D. (1995). *Brave new schools: Challenging cultural illiteracy through global learning networks.* New York: St. Martin's Press.

Davison, C. (1999). Missing the mark: The problem with benchmarking ESL in Australian schools. *Prospect, 14*(2), 66–76.

Davison, C. (2001). Identity and ideology: The problem of defining and defending ESL-ness. In B. Mohan, C. Leung, and C. Davison (eds.) *English as a second language in the mainstream: Teaching, learning and identity* (pp. 71–90). Harlow: Longman Pearson.

Davison, C. (2004). The contradictory culture of classroom-based assessment: Teacher assessment practices in senior secondary English, *Language Testing*, 20 (3), 304–33.

DECCD, Executive Support Branch. (1997). *Does information technology improve teaching and learning*? Hobart: Department of Education, Culture and Community Development Tasmania.

Department of Education, Science and Training (DESST). (2002). *Raising the standards*. Canberra: Department of Education, Science and Training.

Eisner, E. (1988). The ecology of school improvement. *Educational Leadership, 45*(5), 24–9.

Eraut, M. (1986). Inservice teacher education. In M. J. Dunkin (ed.) *The international encyclopedia of teaching and teacher education* (pp. 730–43). Oxford: Pergamon Press.

Fullan, M. G. (2003). *Change forces with a vengeance*. London: Routledge Falmer Press.

Fullan, M. and Hargreaves, A. (1992). Teacher development and educational change. In M. Fullan and A. Hargreaves (eds.) *Teacher development and educational change*. London: Falmer Press.

Garrison, R. (2000). Theoretical challenges for distance education in the 21st century: A shift from structural to transactional issues. *International Review of Research in Open and Distance Learning*, 1(1), 1–17.

Geoghegan, W. H. (1994, July 18–20). Whatever happened to instructional technology? [Electronic version]. Paper presented at the 22nd Annual Conference of the International Business Schools Computing Association. Baltimore, MD. Retrieved November 2000, from http://franklin.scale.uiuc.edu/scale/links/library/geoghegan/wpi.html.

Gipps, C. (1994). *Beyond testing: Towards a theory of educational measurement*. London: Falmer Press.

Halliday, M. A. K. (1993). Towards a language-based theory of learning. *Linguistics and Education, 5*, 93–116.

Hargreaves, A. (1994). *Changing teachers, changing times*. London: Cassell.

Hawkins, M. and Irujo, S. (1993). Paradoxes and perspectives. *TESOL Matters* (April–May), 13.

Hung, D.W.L. and Chen, D. (2001). Situated cognition, Vygotskian thought and learning from the communities of practice perspective: Implications for the design of web-based e-learning. *Education Media International, 38*, 1, 4–12.

Ingvarson, L. and MacKenzie, D. (1988). Factors affecting the impact of inservice courses for teachers: Implications for policy. *Teaching and Teacher Education, 4*(2), 139–55.

Jones, L. and Moore, R. (1993). Education, competence and the control of expertise. *British Journal of the Sociology of Education, 14*(4), 385–97.

Kearsely, G. (2000). *Online education: Learning and teaching in cyberspace*. Belmont, CA: Wadsworth.

Kourtis-Kazoullis, V. and Skourtou, E. (in press). The Internet and English language learning: Opening up spaces for constructivist and transformative pedagogy through sister-class networks. In J. Cummins and C. Davison (eds.) *International Handbook of English Language Teaching* (Vol. 2). Norwell, MA: Springer.

Lankshear, C. and Snyder, I., with Green, B. (2000). *Teachers and techno-literacy. Managing literacy, technology and learning in schools*. St Leonards, Australia: Allen and Unwin.

Lave, J. and Wenger, E. (1991). *Situated learning, legitimate peripheral participation*. Cambridge: Cambridge University Press.

Lawnham, P. (2002, August 2–3). Study warns teachers on IT abilities. *The Weekend Australian*, p. 11.

Little, J. (1984). Seductive images and organisational realities in professional development. *Teachers College Record, 86*(1), 84–102.

Logan, L. and Sachs, J. (1988). Inservice education in Queensland: Some lessons. *The South Pacific Journal of Teacher Education, 16*(1), 63–9.

National Board of Employment Education and Training. (1998). *Information technology: Sink or swim*. Canberra: Commonwealth of Australia.

NCATE/ISTE 2000. Retrieved on July 2, 2003, from http//:wwww.iste.com.

Nunan, D. (in press). *Standards-based approaches to the evaluation of ESL instruction*. In J. Cummins and C. Davison (eds.) *International Handbook of English Language Teaching* (Vol. 1). Norwell, MA: Springer.

Reynolds, M. and Salters, M. (1995). Models of competence and teacher training. *Cambridge Journal of Education, 25*(3), 349–59.

Richey, R.C., Fields, D.C., and Foxon, M. (with Roberts, R.C., Spannaus, T., and Spector, J. M.) (2001). *Instructional design competencies: The standards* (3rd ed.). Syracuse, NY: ERIC Clearinghouse on Information and Technology. ED 453 803.

Rosenberg, M. J. (2001). *E-learning: Strategies for delivering knowledge in the digital age*. New York: McGraw Hill.

Scheffler, F. L., and Logan, J. P. (1999). Computer technology in schools: What teachers should know and be able to do. *Journal of Research on Computing in Education, 31*(3), 305–326.

Strong, R. and Hogan, S. (1994). *TESOL teacher competencies document*. Paper presented at the 28th Annual TESOL Convention, Baltimore, MD.

Taylor J.C. (1994). Technology, distance education and the tyranny of proximity. *Higher Education Management, 6*(2), 179–90.

TESOL (2002). *TESOL standards for teachers of adult learners*. Alexandria, VA: TESOL.

The Independent ICT in Schools Commission. (1997). *Information and communications technology in UK schools: An independent inquiry*. London: The Independent ICT in Schools Commission.

Vygotsky, L. (1978). *Mind in society: The development of higher psychological processes*. Cambridge, MA: Harvard University Press.

Victorian Learning Technologies Capabilities Guide (2000). Retrieved July 10, 2003, from http://www.acce.edu.au/tltc/b-app5.asp.

Walker, J. and Evers, C. (1984). Conflicts in consciousness: Imperatives can lead in knots in thinking. In R. Halkes and J. Olsen (eds.) *Teacher thinking: A new perspective on persistent problems in education*. Lisse: Swets and Zeitlanger.

Wiliam, D. (2001). An overview of the relationship between assessment and the curriculum. In D. Scott (ed.) *Curriculum and assessment* (pp. 165–81). Westport, CO: Ablex Publishing.

Yaxley, B. (1991). Teaching, teacher thinking and teacher development. In P. Hughes (ed.) *Teachers' professional development* (pp. 82–94). Melbourne: Australian Council of Educational Research.

Zeichner, K. (1983). Alternative paradigms of teacher education. *Journal of Teacher Education, 34*(3), 3–9.

Index